D1358054

Ratio Correlation

Ratio Correlation

A Manual for Students of Petrology and Geochemistry

Felix Chayes

THE UNIVERSITY OF CHICAGO PRESS
Chicago and London

ISBN: 0226-10218-1 (clothbound); 0226-10220-3 (paperbound)
Library of Congress Catalog Card Number: 71-146110

The University of Chicago Press, Chicago 60637
The University of Chicago Press, Ltd., London

Published 1971
Printed in the United States of America

Contents

Preface

Ratios and ratio correlation are of central importance to petrography, petrology, and geochemistry, as they are, indeed, to all natural science. The parts per hundred, thousand, or million in which we report our raw data are proportions, already ratios of a rather complicated sort, and from them we generate an endless stream of new ratios which, in the context of one or other specific problem, may seem and sometimes are more germane, informative, or meaningful than the originals.

The process of ratio formation often imposes on its end products interrelations very different from those characterizing the raw variables. This is no reason for not using ratios, but it does suggest that a reasonably thorough knowledge of the numerical consequences of ratio formation should be a prerequisite for those planning careers of research in geochemistry. This book is an attempt to provide such background material in a form suitable for formal course work or self-instruction. The material discussed and the general path of the discussion are described in the first chapter. Here I should like to explain, and perhaps justify, the rather unorthodox manner of presentation.

The overall development is one in which the part of formal statistics, though crucial, is still small. What is called for, over and over again, is close reasoning together with a rather generous exercise of elementary algebra. A working knowledge of elementary statistics, a good background in petrography, and an acute interest in the ratio correlations so common in petrology are presumed. The book is intended as a text, not a reference. With few exceptions, chiefly statistical in nature, the argument is self-contained. Much of the work has appeared previously in the journal articles cited in the references, but I believe the treatment given here is superior for pedagogic purposes; and, in any event, it has the advantage of a uniform notation. The problems, like the text, are intended to teach petrographers and geochemists, not to test mathematicians: the answers are usually given, and there are numerous hints about how to reach them.

I have tried throughout to avoid coming before my fellow petrographers as an expert in statistics. Where purely statistical results of any complexity are required they are stated without proof. The reader who glances hastily through the book will notice that it also contains very little of what would ordinarily be called petrography. What it does contain is a careful and, I believe, unique analysis of the linkage between statistical testing and petrographic ratio correlation. From the proposed null model, the definitions, and the approximation technique outlined in chapter 1, for instance, any alert petrographer or geochemist should long ago have been able to pass directly to the testing procedure of chapter 6, for the null values used in chapter 6 are completely implicit in the material of chapter 1, none of which is novel. Important and fruitful as such a passage would have been for petrography, however, no one has made it. I have therefore not hesitated to state it in detail probably sufficient to dismay, for rather different reasons, both the conventional petrologist and the professional mathematician. In later chapters, similarly, the stress is neither on statistics nor on petrography, but on the taxing and perhaps sometimes tedious business of bringing the one to bear on the other. I hasten to add that the amount of bearing obtained to date is rather modest, requiring us, for instance, to stop talking a great deal of nonsense about petrographic variation diagrams but not as yet permitting us to make much sense of them. Still, the ice is broken; soon we shall all have to learn to swim or face the consequences.

The book has grown, slowly and rather painfully, out of a series of lectures given while I was a visiting professor in the geology department of Northwestern University in the winter quarter of 1966 and again, in abbreviated form, to students in the geology department of Harpur College, State University of New York, in the spring of 1967. I am indebted to faculty and students of both departments, and especially to W. C. Krumbein and W. R. James, for much stimulating discussion and criticism. Whatever is closely reasoned and clearly stated in this book owes much to the criticism and advice of J. M. Cameron and W. Kruskal. The mistakes are mine; I would appreciate being informed of them by readers.

Ratio Correlation

1 Ratios, Relatedness, and Correlation in Petrography

1. Ratios in Petrography

Descriptive petrography is a web of ratio correlations. The basic data of our subject are percentages, for it is only by expressing the results of chemical or modal analyses as proportions that we can meaningfully compare the compositions of groups of specimens. This is so obvious that we never think of it—even in reporting single analyses it is mandatory to record the results as percentages. Yet the percentage is already a rather complicated ratio; and in apprais-ing the relation between any pair of variables in a group of analyses, we cannot assume either that lack of correlation implies lack of association or that numerically strong correlation implies signifi-cant association. The fact that the sum of the variables in each item is a constant common to all items destroys the potential indepen-dence of variances and covariances upon which elementary correla-tion theory rests. The principal algebraic consequence of this "closure restraint" is to make the sum of the covariances of each variable exactly equal to its variance, and opposite in sign. Even in the absence of any substantive relation between the variables there is thus a bias toward negative correlation; the use of departures of r from a parent $\rho = 0$ as a measure of relatedness is clearly suspect.

In most petrographic variation diagrams, however, the original per-centages are not used directly. Rather, the coordinates actually plotted are obtained by one of the following devices:

a. Reduction of some subset of the original variables to a new set of proportions, as when we "project" from the multivariate sample space into the 2- or 3-space of the ternary or quater-nary diagram.

b. Formation of sums which are subsets, with or without common elements, of the original variables, as when we plot MgO against $FeO + Fe_2O_3$, or FeO against $FeO + Fe_2O_3 + MgO$.

c. Formation of ratios, with or without common elements, either between the original variables or between sums which are subsets of the original variables, as when we plot $MgO/(MgO + FeO + Fe_2O_3)$ against SiO_2, or $MgO/(MgO + FeO)$ against $Na_2O/(Na_2O + K_2O)$.

The first of these procedures simply strengthens the closure effect. In any ternary closed array, for instance, all three correlations can be computed exactly from the variances. Further, even if all three of the correlations between the original variables are strongly positive, at least two of the three correlations between their ternary equivalents will be negative. The larger the subset, of course, the less drastic the effect of the reduction on correlation. In general, however, it must always tend to increase the already considerable bias toward negative correlation between major variables. (Paradoxically, it may also introduce a bias toward *positive* correlation between minor variables.)

Providing common elements are lacking, the second procedure necessarily has the same general effect as the first. This may be offset somewhat by the introduction of common elements, but the compensation, which is difficult to evaluate, is by no means exact, and may not be desirable.

Although, as is shown later, the third procedure eliminates the initial closure effect, the means, variances, and covariances of the original data, which were in fact subject to the closure effect, clearly determine the correlations between the ratios. Further, the ratio correlations are themselves subject to curious restrictions which have been the subject of intermittent discussion since the appearance of Pearson's famous essay on "spurious" correlation in 1896. A generalized form of Pearson's definition of spurious correlation, central to the work of this book, is proposed in the next section. Here, however, I need point out only that there is no single or readily apparent null value against which to test the significance of the correlation between ratios formed from percentages.

The conversion of oxide percentages to molar or atomic amounts shifts the variances but has no effect on the correlations. The reduction of these new variables to molar or atomic percentages merely imposes a new closure restraint. Thus, whether we use the raw data or any of a wide variety of conventional transformations, we are in deep trouble whenever we are called upon for an objective numerical characterization, or test, of the strength of a petrographic association. That we ordinarily make no attempt to remedy the difficulty—and indeed, scarcely acknowledge its existence—in no way lessens its force.

Consider, for instance, the scatter diagrams shown in figure 1.1. Most observers would agree that in each there is a strong tendency toward inverse variation between the variables concerned, and in fact the correlation coefficients computed from the data are −0.78

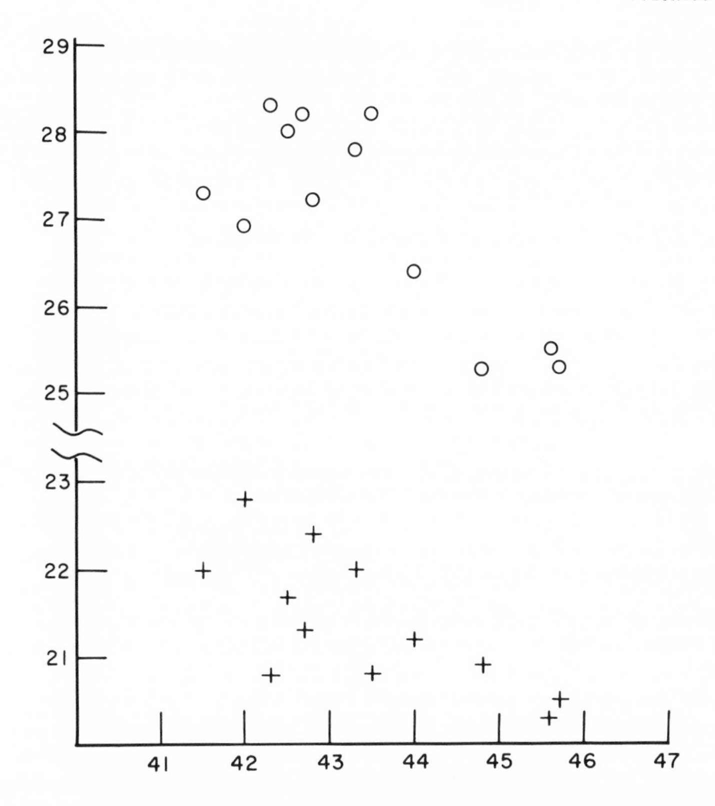

Fig. 1.1 Scatter diagrams illustrating tendency toward negative correlation between the major variables of a closed array.

and −0.73, easily large enough to warrant rejection of the usual null hypothesis that the parent variables are uncorrelated. Yet the only correlation which makes sense for this particular set of observations is that which arises between objects attempting to occupy the same space at the same time. The data are modal analyses of thin sections cut from a single block of granite which, because of its apparent uniformity and homogeneity, was chosen as a reference material for interlaboratory calibration. The modes are given in table 6.1 and examined in detail in chapter 6; they are used again in the exercises of chapter 7.

Now there is no question of the reality of the inverse variations suggested by the scatter diagrams; the competition for space can easily—in fact, must—generate such relations. The troublesome point is just that although they are physically unavoidable they may be of little substantive interest. How objects fill space is a problem for mathematicians and, perhaps, physicists, but the study of petrographically meaningful associations between the variables of a closed array can hardly begin until some reasonable and systematic allowance can be made for the effect of the closure restraint on variances and covariances.

2. Petrographic Relation and Numerical Correlation

If we are to test the significance of association between proportions —or between linear combinations or ratios of proportions—we must have some clearly defined null state, the correlations in which we accept as indicative of a lack of substantively meaningful association. The optimum situation would be one in which we could accept zero correlation as a criterion of unrelatedness. This is, in fact, the assumption underlying most elementary correlation analysis, and one of its major advantages is the resulting analogy between numerical and graphical notions of relatedness. If the points on a scatter diagram tend to follow some linear or moderately curvilinear trend, the numerical correlation will be strong; if they are widely and unsystematically scattered, the numerical correlation will be weak.

In textbooks on the subject one is always, and quite properly, reminded of the limitations of this analogy, the commonly cited exceptions being examples in which weak correlation leads to a clearly erroneous inference of unrelatedness. The correlation between an angle and its sine, for instance, is zero, yet there is a perfectly systematic relation between these two variables. In the work described here the analogy between numerical and graphical notions of association fails for precisely the opposite reason, namely, that even rather strong correlation may lead to erroneous inferences of relatedness between the variables involved. The intuitive demand that zero correlation be used as a criterion of unrelatedness will obviously have to be met in a rather indirect fashion.

Pearson long ago gave an approximation for the correlation between ratios as a function of the means, variances, and covariances of the "absolute" variables, that is, the numerators and denominators of the ratios. He pointed out that even if these were completely uncorrelated, there would nevertheless be correlation between many pairs of ratios formed from them, and such correlation between two ratios with common denominator he termed "spurious". Extension of the concept to correlation between other simple ratios—that is, to ratios with common numerator, ratios of which the numerator of one is the denominator of the other, and so on—is implicit in his essay. The pejorative overtone of the word "spurious" in this context is unfortunate, for the underlying concept is extremely useful.

The data of descriptive petrography are rarely simple ratios of the type described by Pearson, their numerators and denominators usually being linear combinations of the raw values, with or without common elements. The argument may be readily extended to these more complicated forms, however; and in this book just such an extension is proposed. In this extension the criterion of "unrelatedness," the null value against which an observed correlation is tested, will be precisely the correlation termed "spurious" by Pearson—that is, the correlation that would be found if the parent variables from which the ratios were formed were uncorrelated. The null state is then one in which, prior to the transformation generating the parameters of immediate interest, the parent variables are indeed uncorrelated. The null value, against which a sample correlation is to be tested, however, is not the zero characteristic of the untransformed variables but the correlation found after, *and thus generated by*, the transformation. This definition of unrelatedness has strong intuitive appeal but is not unexceptionable.

As we shall see, there are important petrographic situations in which it cannot be applied at all, for the reason that the required untransformed or "open" parent evidently does not exist. In chapter 8 I shall discuss two auxiliary transformations whose practical application is based on the assumption that this prior null state is indeed numerically realizable for the variables in which rock compositions are ordinarily expressed—namely, the weight or mole percentages of constituent oxides. For a very large class of such data—the suites of analyses of subalkaline rocks conventionally portrayed in Harker or Niggli variation diagrams—it is now known that this assumption is usually unwarranted. Statistical appraisal of such data must accordingly be preceded by one of two stratagems. Either the null model must be modified, presumably in the direction of relaxing the requirement for independence of all open parent variables, or the raw data must be transformed to variables for which the unmodified null model is numerically realizable. Chapter 9 contains a brief algebraic exploration of the first alternative and a rather extended practical discussion of the second.

3. Notation

We shall be concerned throughout with estimating means, variances, and covariances of variables (V, Q, Y) which are linear combinations of underlying theoretical variables (X), and in subsequent chapters specifically reserve Y to denote proportions, that is, $Y_i = X_i/\Sigma X$.

From a hypothetical random vector,

$$\mathbf{X} = [X_i, X_j, \ldots, X_m], \tag{1.1}$$

whose non-negative, uncorrelated elements have means $\boldsymbol{\mu} = [\mu_i, \mu_j, \ldots, \mu_m]$ and variances $\boldsymbol{\sigma} = [\sigma_i^2, \sigma_j^2, \ldots, \sigma_m^2]$ we form a random vector, **Y**, each of whose elements is some linear combination of the X's, namely,

$$Y_k = \boldsymbol{\lambda}'_k \mathbf{X}, \qquad k = 1, n \tag{1.2}$$

the combinations of major interest being ratios of sums of two or more elements of **X**. (It may sometimes happen that Y_i is in some obvious sense identifiable with X_i for all i so that a single set of subscripts suffices for both vectors, but this is not necessary.)

Using the parameters of **X** and the usual rules for calculating expectations,[1] we then find

$$E(Y_k) \cong E(\boldsymbol{\lambda}'_k \mathbf{X}), \tag{1.3}$$

where E( ) denotes the expected value of the bracketed term. If the combination in question involves a sum, difference or product of X's, equation (1.3) will be exact. If it involves a ratio of X's, however, equation (1.3) will be a linear approximation. Since our principal concern will be with ratios, we begin at this point to use $\cong$ in place of $=$; the reader should remember that if Y_k contains no ratios of X's, $E(Y_k)$ is exact.

The deviations of any pair of Y's, say Y_k and Y_p, from their expected values are

$$\begin{aligned} \Delta_k &= Y_k - E(Y_k) \cong \boldsymbol{\lambda}'_k \mathbf{X} - E(\boldsymbol{\lambda}'_k \mathbf{X}), \\ \Delta_p &= Y_p - E(Y_p) \cong \boldsymbol{\lambda}'_p \mathbf{X} - E(\boldsymbol{\lambda}'_p \mathbf{X}). \end{aligned} \tag{1.4}$$

From these we form the quantities Δ_k^2, Δ_p^2, $\Delta_k\Delta_p$, and again take expectations, to find

$$\begin{aligned} \text{var}\,(Y_k) &\cong E(\Delta_k^2), \\ \text{var}\,(Y_p) &\cong E(\Delta_p^2), \\ \text{cov}\,(Y_k Y_p) &\cong E(\Delta_k \Delta_p). \end{aligned} \tag{1.5}$$

Thus, finally,

$$\rho_{kp} \cong \text{cov}\,(Y_k Y_p)/\sqrt{\text{var}\,(Y_k) \times \text{var}\,(Y_p)}, \tag{1.6}$$

ρ_{kp} being the correlation to be anticipated under our null hypothesis. Unless an observed correlation r_{ij} differs with statistical significance from ρ_{ij}, we shall argue that it is "spurious" in the sense of

---

[1] For which see, for instance, Feller, pp. 164-87, or Hoel, pp. 135-38.

Pearson, that is, that it is compatible with a complete lack of correlation between the underlying **X**'s. In more modern language, we use ρ_{ij} as a null value against which to test r_{ij}, our hypothesis being that the underlying variables from which the ratios are formed are uncorrelated.

Thus, for the purpose of the test we agree to suppose that our samples are randomly drawn from **Y**, and whether we ultimately abandon this hypothesis depends on whether or not a sample r differs significantly from ρ. Superficially, the operation seems a rather ordinary example of standard small-sample statistics and indeed that is precisely what it is if **X** is known or hypothesized, so that **Y** can be derived prior to the sampling. When this is not so, however, as in most of the work discussed in chapters 5, 6, 7, and 9, the situation is both more complex and less satisfactory. Where **X** is unknown prior to the experiment, we choose its parameters so that the parent **Y** derived from it will have means and variances identical with those of the sample. (For this reason, in notation concerning **Y** we make no symbolic distinction between parent and sample means and variances.) For the sake of reaching a position that permits at least the beginning of an objective appraisal of the correlations we in effect ignore the influence of sampling variation and experimental error on the estimates of variances and means. A less arbitrary treatment, one which would permit testing of the covariances without destroying the usual sample-parent dichotomy of the means and variances, would certainly be preferable; at present none is known.

4. Approximation Technique

A brief résumé of the approximation technique will be useful to many readers. Any denominator which is a sum of two numbers, say, a and b, may be thrown into the form $b^{-1}(1 + a/b)^{-1}$ and expanded, namely,

$$\frac{1}{b}\left(1+\frac{a}{b}\right)^{-1} = \frac{1}{b}\left[1-\frac{a}{b}+\left(\frac{a}{b}\right)^{2}-\ldots\right]. \tag{1.7}$$

We shall assume throughout that $|a/b| < 1$ and approximate the denominator of any $\boldsymbol{\lambda}'\mathbf{X}$ by the sum of terms of order less than 2 in a on the right of equation (1.7), for example,

$$Y = \frac{X_1}{X_2} = \frac{\mu_1+\delta_1}{\mu_2+\delta_2} = \frac{\mu_1+\delta_1}{\mu_2}\left(1+\frac{\delta_2}{\mu_2}\right)^{-1} \cong \left(\frac{\mu_1+\delta_1}{\mu_2}\right)\left(1-\frac{\delta_2}{\mu_2}\right), \tag{1.8}$$

the quality of the approximation falling off rapidly as $|a/b| \to 1$.

Performing the indicated multiplication and dropping the term in $\delta_1 \delta_2$, whose expectation is zero according to our null model, we find

$$Y \cong \frac{1}{\mu_2^2} (\mu_1 + \delta_1)(\mu_2 - \delta_2) \cong \frac{1}{\mu_2^2} (\mu_1 \mu_2 + \mu_2 \delta_1 - \mu_1 \delta_2), \quad (1.9)$$

the right side of which is in fact what would be obtained by evaluating the linear terms of a Taylor series expansion of X_1/X_2 at μ_1, μ_2 (see, for instance, Ku 1965, p. 14). This approximation technique, used repeatedly throughout the book, is a very simple application of the "delta method," for which see appendix A of Goodman and Kruskal (1963) or chapter 10 of Kendall and Stuart (1963).

2 The Simple Ratio Correlations

Here we are concerned with correlation between ratios whose numerators and denominators are individual values of uncorrelated "raw" variables, for example, $Q_i = X_k/X_n$. The approximate parent correlation between any pair of such ratios may be obtained as a degenerate form of the Pearson general formula,

$$\rho_{ij} \cong \frac{\rho_{13}C_1C_3 - \rho_{14}C_1C_4 - \rho_{23}C_2C_3 + \rho_{24}C_2C_4}{(C_1^2 + C_2^2 + 2C_1C_2\rho_{12})^{1/2}\,(C_3^2 + C_4^2 + 2C_3C_4\rho_{34})^{1/2}}, \tag{2.1}$$

where $Q_i = X_1/X_2, Q_j = X_3/X_4$, and $C_k = \sigma_k/\mu_k$. By "degenerate" we mean here either that the ratios have one term in common, as, for instance, if $X_2 = X_4$, or that, in addition, one of the noncommon terms is a constant, as, for instance, if $X_2 = X_3$ and $X_4 = 1$. There are, in all, five distinguishable degenerate forms—namely, correlation of a ratio with its numerator or denominator, correlation between two ratios with common numerator or common denominator, and correlation between two ratios the numerator of one of which is the denominator of the other.

1. Null Correlation of a Ratio with Its Numerator ($Q_i = X_1/X_2$, $Q_j = X_1$)

By following the procedure described at the close of chapter 1, we

---

The purposes of this chapter are to introduce the novice to the fascinating problems of ratio correlation, to familiarize him with the notation just announced, and to give him practice in the algebraic manipulations used throughout the book. Some of the specific results obtained here are used later, but readers with adequate background will find them easy to derive as needed; such readers may nevertheless be interested in the sampling experiments described in section 3 and in the work of section 5.

have

$$Q_i = \frac{X_1}{X_2} = \frac{\mu_1 + \delta_1}{\mu_2 + \delta_2} = \frac{\mu_1 + \delta_1}{\mu_2}\left(1 + \frac{\delta_2}{\mu_2}\right)^{-1} \cong \frac{\mu_1 + \delta_1}{\mu_2}\left(1 - \frac{\delta_2}{\mu_2}\right),$$

or

$$Q_i \cong \frac{\mu_1}{\mu_2} + \frac{\delta_1}{\mu_2} - \frac{\mu_1}{\mu_2^2}\delta_2. \qquad (2.2)$$

Taking expectations on both sides of equation (2. 2),

$$E(Q_i) \cong \mu_1/\mu_2. \qquad (2.3)$$

Subtracting equation (2. 3) from (2. 2), we have, for the deviation of Q_i from its mean,

$$\Delta_i = Q_i - E(Q_i) \cong \frac{1}{\mu_2^2}(\mu_2\delta_1 - \mu_1\delta_2), \qquad (2.4)$$

and, for its square,

$$\Delta_i{}^2 \cong \frac{1}{\mu_2^4}(\mu_2^2\delta_1^2 + \mu_1^2\delta_2^2 - 2\mu_1\mu_2\delta_1\delta_2). \qquad (2.5)$$

Taking expectations on both sides of equation (2. 5), the expected variance of Q_i is

$$\sigma_i{}^2 \cong \frac{1}{\mu_2^4}(\mu_2^2\sigma_1^2 + \mu_1^2\sigma_2^2), \qquad (2.6)$$

since, according to our null model, $E(\delta_1\delta_2) = 0$.

The "transformation" of X_1 is simply a change of name, so that $E(Q_j) = \mu_1$, $\Delta_j = \delta_1$, and $\sigma_j{}^2 = \sigma_1^2$. Hence

$$\Delta_i\Delta_j = \Delta_i\delta_1 \cong \frac{1}{\mu_2^2}(\mu_2\delta_1^2 - \mu_1\delta_1\delta_2). \qquad (2.7)$$

Taking expectations on both sides of equation (2. 7), the expected covariance of Q_i with Q_j is

$$\sigma_{ij} \equiv E(\Delta_i\Delta_j) \cong \sigma_1^2/\mu_2, \qquad (2.8)$$

again under the condition that $E(\delta_1\delta_2) = 0$.

Finally, the null correlation between Q_i and Q_j, if X_1 and X_2 are uncorrelated, is to a linear approximation,

$$\rho_{ij} = \frac{\sigma_{ij}}{(\sigma_i{}^2\sigma_j{}^2)^{1/2}} \cong \frac{\dfrac{\sigma_1^2}{\mu_2}}{\sqrt{\dfrac{\sigma_1^2}{\mu_2^4}(\mu_2^2\sigma_1^2 + \mu_1^2\sigma_2^2)}} = \frac{\mu_2\sigma_1}{\sqrt{\mu_2^2\sigma_1^2 + \mu_1^2\sigma_2^2}} \cdot \quad (2.9)$$

Although sometimes left implicit in the remainder of the book, except in section 5 of this chapter and section 3 of chapter 9, the zero covariance property of our null model is critical in every passage from algebraic formulation to statistical expectation—as here, for instance, in passing from equation (2. 5) to (2. 6) and from equation (2. 7) to (2. 8). This is the first symbolic formulation of the key relation that in general $E(\Delta_i\Delta_j) \neq 0$ even though $E(\delta_i\delta_j) = 0$. In a nutshell, that is what most of the book is about; covariances between ratios or other linear combinations of variables are governed by the means and variances of the parent variables if the latter are uncorrelated.

Exercises

2. 1 By setting $X_3 = X_1$ and $X_4 = 1$, show that equation (2. 1) reduces to $C_1/(C_1^2 + C_2^2)^{1/2}$.

2. 2 Replacing C_1 and C_2 by their definitions, show that the result of exercise 2. 1 is equivalent to equation (2. 9).

2. Other Forms of Simple Ratio Correlation

Proceeding in exactly analogous fashion, the reader wishing to familiarize himself with the notation and the approximation procedure used throughout the book will find it useful to derive the symbolic form of the null correlation for each of the other identifiable cases of simple ratio correlation. All are shown in table 2. 1, where, for completeness, the result obtained in the preceding section is also included. The symmetry of the results for the correlation of a ratio with its numerator (line 1) and denominator (line 2) is evident; the denominators are identical, the numerators are related by a rotation of subscripts, the ratio varies directly with its numerator and inversely with its denominator. Perhaps more surprising, correlation between ratios with common denominator (line 3) is identical with that between ratios with common numerator (line 4), and differs only in sign from that between ratios of which the numerator of one is the denominator of the other (line 5). The correlation between ratios with common denominator was specifically designated as "spurious" by Pearson and is so labeled in most of the scattered literature of the subject. But his argument clearly applies to all five forms.

Table 2.1 Approximate Null Correlations between Ratios Formed of Uncorrelated Variables

Variables	Approximate Null Correlation
$X_1/X_2; X_1$	$\mu_2\sigma_1/(\mu_2^2\sigma_1^2 + \mu_1^2\sigma_2^2)^{1/2}$
$X_1/X_2; X_2$	$-\mu_1\sigma_2/(\mu_2^2\sigma_1^2 + \mu_1^2\sigma_2^2)^{1/2}$
$X_1/X_2; X_3/X_2$	$\mu_1\mu_3\sigma_2^2/\{(\mu_2^2\sigma_1^2 + \mu_1^2\sigma_2^2)(\mu_2^2\sigma_3^2 + \mu_3^2\sigma_2^2)\}^{1/2}$
$X_2/X_1; X_2/X_3$	$\mu_1\mu_3\sigma_2^2/\{(\mu_2^2\sigma_1^2 + \mu_1^2\sigma_2^2)(\mu_2^2\sigma_3^2 + \mu_3^2\sigma_2^2)\}^{1/2}$
$X_1/X_2; X_2/X_3$	$-\mu_1\mu_3\sigma_2^2/\{(\mu_2^2\sigma_1^2 + \mu_1^2\sigma_2^2)(\mu_2^2\sigma_3^2 + \mu_3^2\sigma_2^2)\}^{1/2}$
$X_1/X_2; X_3/X_4$	0

We ignore here the circumstance that the geochemical variables of which ratios are formed are often themselves already ratios of a rather intractable sort, namely, proportions. It is shown in section 4 of this chapter that this occasions no difficulty when a null value for the correlation of two ratios is sought. When desired null correlation is of a ratio with one of its own terms, however, whether numerator or denominator, the fact that the basic variables are proportions materially complicates the derivation. We shall return briefly to this subject in exercise 3.3.

Exercises

2.3 Divide numerator and denominator of the correlation in the third line of table 2.1 by $\mu_1\mu_2^2\mu_3$ to obtain $C_2^2/\{(C_1^2 + C_2^2)(C_2^2 + C_3^2)\}^{1/2}$ and show that this is the result found from equation (2.1) if $X_4 = X_2$.

2.4 Obtain the correlation in the fourth line of table 2.1 from equation (2.1) by interchanging subscripts 1 and 2, and then replacing 3 by 2 and 4 by 3.

2.5 Show that the correlation in the fifth line of table 2.1 may also be obtained as a special case of equation (2.1).

2.6 The square of the coefficient of variation, or relative standard deviation, is called the relative mean-square error. Divide both sides of equation (2.6) by $(\mu_1/\mu_2)^2$ to show that $C_i{}^2 = C_1^2 + C_2^2$, that is, the relative mean-square error of a ratio is the sum of the relative mean-square errors of its numerator and denominator.

3. Some Sampling Experiments in Ratio Correlation

The easy, straightforward algebra of ratio correlation is just about exhausted by the discussion of the preceding sections, and it leaves many important questions unanswered. Chief among these are the effects of dispersion and correlation on the adequacy of the approximations. Difficult to evaluate by a priori analysis, these matters are readily subjected to experimental examination; results obtained in this fashion of course lack generality, but it is now a simple matter to run a simulation experiment on variables characterized by any set of means, variances, and covariances that may be of practical interest. In the simulations reported here, only the means and variances of the parent variables are adjusted; the parent covariances are all zero, and those generated in reasonably large numerical samplings are always very small.

Everyone agrees that good approximations of ratio correlations by the methods described in the preceding section may be obtained only if the coefficients of variation—the ratios $C = \sigma/\mu$ for all variables—are small, but no one says just what is meant by "small" in this connection. Table 2. 2 reports results for five experiments bearing on this question. Entries in columns 4-8 of the table were obtained in the following way:

1. To generate 2000 sets of values for (X_1, X_2, X_3) a quasi-normal deviate of zero mean and unit variance was multiplied by an assigned constant, σ, and added to a second assigned constant, μ, to yield a sample from a normal population with mean μ and standard deviation σ. (In the first experiment the ratio $C = \sigma/\mu$ had the value shown as the caption of column 4, namely 0. 05; in the second, C was 0. 10, etc.)

2. For $k = 1, 2000$, (X_{1k}, X_{2k}, X_{3k}) were transformed to (Q_i, Q_j) in the ways shown by the successive row captions at the left of the table, and the correlation between the pair of Q's in each row was computed. (In the first step of the first experiment, it was found that $r = 0.7061$ between $Q_i = X_1/X_2$ and $Q_j = X_1$, in the second that $r = -0.7009$ between $Q_i = X_1/X_2$ and $Q_j = X_2$, etc.)

The third column of the table, headed "Null," contains values for ρ calculated directly from the fomulae given in table 2. 1. (Since $\mu_1 = \mu_2 = \mu_3$ and $\sigma_1 = \sigma_2 = \sigma_3$ throughout, the first null value is simply $1/\sqrt{2}$, the second $-1/\sqrt{2}$, the third and fourth $1/2$, and the fifth $-1/2$.)

For C larger than 0. 15 the differences between predicted and observed correlations may be large, but for $C \leq 0.15$ the approximations are quite good. The situation is little changed if the means are varied, providing C is held constant. Variation in C, depending on its relation to mean values, may either enlarge or reduce the differences between specific predicted and observed correlations.

It is important to realize that large differences do not imply that the relations between ratios are somehow independent of the properties of the variables from which the ratios are formed. Even large differences between predicted and observed correlations may reflect

Table 2.2 Experimentally Determined Ratio Correlations for Variables with Common Mean and Variance

Ratio		Null	Coefficient of Variation of X_1, X_2, and X_3				
Q_i	Q_j	Correlation	0.05	0.10	0.15	0.20	0.25
X_1/X_2	X_1	0.7071	0.7061	0.6750	0.6669	0.6395	0.6193
X_1/X_2	X_2	−0.7071	−0.7009	−0.6955	−0.7103	−0.7007	−0.7165
X_1/X_2	X_3/X_2	0.5000	0.5030	0.4963	0.5234	0.5588	0.6207
X_2/X_1	X_2/X_3	0.5000	0.4972	0.4664	0.4440	0.3956	0.3014
X_1/X_2	X_2/X_3	−0.5000	−0.4973	−0.4745	−0.4719	−0.4323	−0.3557

nothing but the inadequacy of the approximation procedure. Before the development of numerical simulation, this inadequacy was of course critical. As pointed out above, however, it is now possible to solve almost any specific problem of this sort experimentally. If a large enough simulation experiment shows that the correlations observed between ratios could (or could not) reasonably have been generated by formation of the ratios from the (uncorrelated) "absolute" variables in question, the resulting statistical inference is for all practical purposes quite as strong as if the null value had been found independently of numerical experiment. An approximation technique able to tolerate large and variable coefficients of variation would certainly be preferable, but is no longer indispensable for the solution of specific problems.

Difficulties encountered because of excessive size, or variation, of the coefficients of variation of the "raw" or "absolute" variables from which the ratios are formed are thus essentially practical. The principal issue in any discussion of the *covariances* of these variables, however, is of a much more fundamental nature, for the correlation between any pair of ratios formed from any three variables is completely implicit in the means, variances, and covariances of those variables. The means and variances are already parameters of our null model, and if we now include the covariances as well, correlations between ratios lose most of their statistical character. If the parameters of the model are not all identical with those of the raw variables we may draw statistical inferences about whether or not a sample of ratios could have been drawn from a parent such as that specified by the model. Under these circumstances a test of the null hypothesis—the hypothesis that the sample could indeed have been randomly drawn from a parent conforming to such a model—is worth making.

If, on the other hand, the parameters of the model—means, variances, *and* covariances—are identical with those of the raw variables from which the ratios are formed, samples of ratios are indeed drawn from a parent conforming to the model, and nothing is to be gained by a test. In fact, the only statistically significant results of such a test would uniformly lead to "errors of the first kind," that is, rejections of a null hypothesis which is in fact true.[1]

It is a commonplace that the inclusion of nonzero covariances in a null model of the sort proposed in chapter 1 vastly complicates and extends the algebra by means of which the model is put to work. The

---

[1] All of this presumes that, and is applicable only when, the variables from which the ratios are to be formed are directly measurable, and the formation of the ratios is at the option of the experimenter. This is very often but by no means always the case, as we shall see in later chapters.

discussion of this section shows that it may also introduce conceptual difficulties whose discussion, like that of their algebraic counterparts, is rather out of place in an elementary manual. Except for brief digressions in the concluding section of this chapter and in section 3 of chapter 9, therefore, we shall specify, as announced in chapter 1, that the variables of the null model are uncorrelated, that is, that their covariances are zero.

The possible applicability of such a simple model to the real world of often highly correlated variables is perhaps more easily argued by example than by precept. Table 2.3 shows the sample statistics for the proportions of various isotopes of Pb, as determined in some 350 samples analyzed in the Geophysics Laboratory of the University of Toronto. The sample correlations between the ratios 206/204 and 207/204, 206/204 and 208/204, 207/204 and 208/204, are shown in table 2.4. The right column of the table shows analogous correlations obtained in sampling experiments (N = 1,000), with variables having the means and standard deviations shown in the first two lines of table 2.3, but zero covariances. The null and sample correlations for 206/204 and 207/204 differ by so much that the null hypothesis may be discarded with overwhelming assurance; the sample correlation cannot have been generated by the formation of ratios from uncorrelated variables. But a difference like that between the null and sample correlations of 206/204 with 208/204 is much more marginal; deprived of the sanctity usually accorded

Table 2.3 Sample Statistics Computed from 350 Determinations of Proportions of Pb Isotopes

Isotope	204	206	207	208
Mean	1.4155	24.3686	22.0850	52.1278
Standard deviation	0.1311	3.0018	1.2696	2.0437

	Correlations		
Isotope	206	207	208
204	−0.8794	0.9687	0.6252
206		−0.7986	−0.9148
207			0.4900

Note: From the data of Russell and Farquhar, appendix 1.

Table 2.4 Sample and Experimentally Determined Null Correlations for Three Ratio Correlations

Correlated Ratios	Observed Value	Null Value (N = 1,000)
(206/204) with (207/204)	0.9660	0.5635
(206/204) with (208/204)	0.7488	0.5632
(207/204) with (208/204)	0.7512	0.7921

Note: Data of table 2.3.

results based on highly instrumented measurements, it would not be regarded as significant at the 0.01 level unless based on a sample of more than 50 specimens. Now in fact, as table 2.3 shows, the proportions of isotopes 204, 206, and 208 are highly correlated; nevertheless, even rather large samples of ratios formed from randomly drawn uncorrelated variables having the means and variances of 204, 206, and 208 might exhibit correlation as strong as that actually observed.

4. Simple Ratios Formed from Proportions

We have so far been concerned with ratios formed directly from the uncorrelated parent variables of the null model proposed in chapter 1. Most of the compositional data of petrography are proportions, and proportions, because of the closure effect, are not, in general, uncorrelated. Is it reasonable to suppose that null values for ratio correlations derived without consideration of the closure restraint will have much relevance for the study of ratios formed from proportions? The answer is that it does seem unreasonable, but is nevertheless often true. (The Pb isotope relations just described are a case in point.)

Suppose that Q_i and Q_j were formed from Y's instead of X's, where, letting $T = \sum_1^m X_n$,

$$Y_k = X_k/T, \qquad 1 \leq k \leq m,$$

so that each Y is a proportion. Then, if Q_i is the ratio of any two Y's, say Y_1 and Y_2,

$$Q_i = \frac{Y_1}{Y_2} = \frac{X_1/T}{X_2/T} = \frac{X_1}{X_2}, \qquad (2.10)$$

and similarly for Q_j if it is a ratio of proportions or linear combinations of proportions, but not otherwise.

Evidently it sometimes does not matter whether we start with proportions or with variables not subject to the closure constraint. The null values for correlation between various simple ratios may indeed be applicable to the data of petrography and geochemistry even though derived from variables of a type rarely encountered in our subject. As a corollary, the formation of ratios from proportions eliminates the closure restraint; more precisely, it replaces the specific constraint inherent in closure by a more general set of constraints inherent in ratio formation. With a single exception, the use of $\rho_{ij} = 0$ as a criterion of unrelatedness is as unrealistic if the data are simple ratios as it is if they are proportions.

The exception is interesting and potentially useful. If $\rho_{km} = 0$ for all (X_k, X_m), $k \neq m$, the numerator of equation (2.1) goes to zero. Thus, if two simple ratios formed from proportions contain no common term, the expected correlation between them is zero. In a well known petrographic diagram, the ratio $Q_i = Na_2O/(Na_2O + K_2O)$ is plotted against $Q_j = MgO/(FeO + MgO)$; $\rho_{ij} = 0$ is, in fact, the appropriate null value against which to test this association.

Exercises

2.7 Show directly that σ_{ij}, and hence ρ_{ij}, is zero if $Q_i = X_1/X_2$, $Q_j = X_3/X_4$.

2.8 In a simulation experiment with $N = 2,000$, variables (X_1, X_2, X_3) were assigned means (50, 25, 50) and standard deviations (2.5, 5.0, 2.5). Sample correlations between these variables are $r_{12} = 0.0464$, $r_{13} = 0.0018$, $r_{23} = -0.0163$. The sample correlation of X_1/X_2 with X_2 is -0.927, that of X_1/X_2 with X_3/X_2 is 0.951, and that of $X_1/(X_1 + X_2)$ with $X_3/(X_2 + X_3)$ is 0.941. Show, either directly or by substitution in the appropriate line of table 2.1, that the expected value of the first of these ratio correlations is -0.970, while that of the second and third is 0.941.

5. Correlation of a Ratio with Its Numerator When Numerator and Denominator Are Correlated ($Q_i = X_1/X_2, Q_j = X_1, \rho_{12} \neq 0$).

Although somewhat off the main path proposed in chapter 1, this case is of considerable practical interest.[2] Since the approximation is

---

[2] This section treats a special topic and may be omitted on a first reading. The argument of section 3 of chapter 9 is essentially a multivariate version of the material covered here through equation (2.15); the discussion that follows (2.15) presumes familiarity with the work of section 2 of chapter 6.

still to be linear, the expressions for Q_i, $E(Q_i)$, and Δ_i are just those already given in section 1, namely,

$$Q_i \cong \frac{\mu_1}{\mu_2} + \frac{1}{\mu_2^2}(\mu_2\delta_1 - \mu_1\delta_2), \tag{2.2}$$

$$E(Q_i) \cong \mu_1/\mu_2, \tag{2.3}$$

and

$$\Delta_i \cong \frac{1}{\mu_2^2}(\mu_2\delta_1 - \mu_1\delta_2). \tag{2.4}$$

In forming $\Delta_i{}^2$ and $\Delta_i\Delta_j$, in which we already retain terms through order 2 in each δ, we must now also include those in $\delta_1\delta_2$, for the expectation of $(\delta_1\delta_2)$ is no longer zero, as it was in section 1. In fact, we have

$$\Delta_i{}^2 = \frac{1}{\mu_2^4}(\mu_2^2\delta_1^2 + \mu_1^2\delta_2^2 - 2\mu_1\mu_2\delta_1\delta_2), \tag{2.11}$$

from which

$$\text{Var}\,(Q_i) = E(\Delta_i{}^2) \cong \frac{1}{\mu_2^4}(\mu_2^2\sigma_1^2 + \mu_1^2\sigma_2^2 - 2\mu_1\mu_2\sigma_1\sigma_2\rho_{12}), \tag{2.12}$$

and

$$\Delta_i\Delta_j = \frac{1}{\mu_2^2}(\mu_2\delta_1^2 - \mu_1\delta_1\delta_2), \tag{2.13}$$

from which

$$\text{Cov}\,(Q_i, Q_j) = E(\Delta_i\Delta_j) \cong \frac{\sigma_1}{\mu_2^2}(\mu_2\sigma_1 - \mu_1\sigma_2\rho_{12}). \tag{2.14}$$

With some rearrangement of terms, the desired null value may accordingly be written

$$\rho_{ij} \cong \frac{\mu_2\sigma_1 - \mu_1\sigma_2\rho_{12}}{(\mu_2^2\sigma_1^2 + \mu_1^2\sigma_2^2 - 2\mu_1\mu_2\sigma_1\sigma_2\rho_{12})^{1/2}}, \tag{2.15}$$

which is like equation (2.9), except for the terms in ρ_{12}. The intuitively reasonable anticipation is surely that a ratio will vary directly with its numerator, and equation (2.9) shows that this will indeed be so, to the level of approximation used in this book, if $\rho_{12} = 0$. Equation (2.15) now shows that the same thing holds if $\rho_{12} < 0$, but that ρ_{ij} may be either positive or negative if $\rho_{12} > 0$. Specifically, ρ_{ij} will then be positive if and only if

$$\frac{\mu_2\sigma_1}{\mu_1\sigma_2} > \rho_{12},$$

or, equivalently,

$$C_1/C_2 > \rho_{12},$$

where, for any variable, $C = \sigma/\mu$ is the coefficient of variation so widely used in geochemistry. Reliance on this particular ratio correlation may thus lead to highly discrepant inferences in small sample work, for the uncertainty attaching to estimates of variance in such samples is very large.

Some recent work on the K and Rb content of submarine basalt provides a rather dramatic example. In 1965 Gast published a diagram in which, on the basis of the analyses shown here in table 2.5, he outlined a field indicating strong negative correlation between K and the ratio K/Rb in some submarine basalts from the Atlantic and Pacific. In 1970 Hart and Nalwalk published a diagram, based on the data shown in table 2.5, indicating strong positive correlation between K and K/Rb in some submarine basalts of the Puerto Rico trench.

Denoting K and Rb by subscripts 1 and 2, respectively, in the Gast data r_{12} is 0.99—the amount of Rb is almost exactly proportional to the amount of K—and C_1/C_2 is 0.60, so the negative correlation between K/Rb and K is hardly much occasion for surprise. In the Hart-Nalwalk data, on the other hand, r_{12} is 0.42, C_1/C_2 is 1.89, and the correlation of K/Rb with K is positive, as it should be.

Note that in *both* sets of data the correlation of K with Rb is positive. The Gast correlation is strong enough so that, unaccompanied by other information, it would justify an inference of positive association between K and Rb; indeed, it is strong enough to suggest an almost functional relationship. The Hart-Nalwalk correlation, on the other hand, is low enough to be compatible with the hypothesis that K and Rb are in fact uncorrelated in the parent population. But it is still not generally realized how vulnerable estimates of correlation are to sample size. The lower bound of the 0.99 confidence interval about the Gast correlation is 0.88; the upper bound of the 0.99 confidence interval about the Hart-Nalwalk correlation is 0.85. (For instructions on calculating confidence intervals about a sample correlation, see, for instance, Snedecor 1956, p. 175.) There does appear to be justification for inferring that different correlations of K with Rb characterize the parents from which these samples were drawn, but as between samples of this size the inference is just

Table 2.5 Parts per Million of K and Rb in Some Submarine Basalts

	Atlantic and Pacific		Puerto Rican Trench	
	K	Rb	K	Rb
	4500	9.50	5600	2.925
	3040	5.61	4250	2.094
	2285	3.69	3082	2.069
	1560	1.14	3056	2.043
	1400	0.98	2555	1.772
	1345	1.42	2435	3.440
	875	0.75	2084	1.389
	641	0.35	2042	1.366
			1560	2.830
			1237	2.368
			937	2.175
			642	1.263
			620	1.259
$\bar{x}$	1955.75	2.93	2315.38	2.076
s	1281.31	3.1987	1448.44	0.6858
C	0.65515	1.09170	0.62557	0.33027
r	0.9876		0.4163	

Note: Atlantic and Pacific, data of Gast (1965); Puerto Rican Trench, data of Hart and Nalwalk (1970).

barely permissible. One would hardly guess this from the correlations between K/Rb and K in the two sets of data; for the Gast data r_{ij} is −0.82 and for the data of Hart and Nalwalk it is +0.86!

That the sample ratio correlations differ in sign is not in itself particularly critical. The data of Hart and Nalwalk are closely compatible with our parent null hypothesis—namely, that $\rho_{12} = 0$; the appropriate null value for the ratio correlation under this hypothesis is 0.88, as compared to an observed value of 0.86. For the Gast data, on the other hand, ρ_{ij} is +0.52 as compared to an observed value of −0. 82, a difference easily significant even with n as small as 8. Nevertheless, in randomly drawn samples of 8 from a parent with C_1 and C_2 as in the Gast data but $\rho_{12} = 0$, that is, a parent in which there was no nonrandom association between K and Rb, the ratio correlation would be negative almost as often as not, and $\rho_{ij} = +0.52$ would be within the upper bound of the 0. 99 confidence interval about any observed negative ratio correlation less than 0. 67 in absolute value.

3 The Part-Whole and Common-Element Correlations

In the first two sections of this chapter we derive two simple results which, because of their considerable importance in practical work, are well known in many fields of natural science. The discussion is something of a digression, since both results are exact and neither involves a ratio; both are useful, however, in clarifying the argument and may also sometimes reduce the algebraic tedium of later work. Readers sufficiently familiar with the part-whole and common-element correlations should pass directly to sections 3 and 4, in which we return to the general subject of ratio correlation.

1. The Part-Whole Correlation ($Q_i = X_1 + X_2, Q_j = X_1$)

From the definition of Q_j it is evident that

$$\Delta_j = \delta_1 \tag{3.1}$$

and

$$\sigma_j^2 = \sigma_1^2. \tag{3.2}$$

For the whole we have

$$Q_i = X_1 + X_2 = \mu_1 + \mu_2 + \delta_1 + \delta_2, \tag{3.3}$$

so that

$$E(Q_i) = \mu_1 + \mu_2. \tag{3.4}$$

Then

$$\Delta_i = Q_i - E(Q_i) = \delta_1 + \delta_2, \tag{3.5}$$

$$\Delta_i^2 = \delta_1^2 + 2\delta_1\delta_2 + \delta_2^2, \tag{3.6}$$

and, in view of the zero covariance of the X's in our null model, the expected value of the variance of Q_i is

$$\sigma_i{}^2 = \sigma_1^2 + \sigma_2^2. \tag{3. 7}$$

The product of the deviations is

$$\Delta_i\Delta_j = \delta_1^2 + \delta_1\delta_2, \tag{3. 8}$$

and, taking expectations on both sides of equation (3. 8),

$$\sigma_{ij} = \sigma_1^2 \tag{3. 9}$$

since $E(\delta_1\delta_2) = 0$. Hence

$$\rho_{ij} = \frac{\sigma_1}{\sqrt{\sigma_1^2 + \sigma_2^2}}. \tag{3. 10}$$

Under our model the null correlation between the part and the whole is the ratio of the standard deviation of the former to that of the latter.

2. The Common-Element Correlation ($Q_i = X_1 + X_2, Q_j = X_1 + X_3$)

For Q_i we have, from the preceding section, that

$$\Delta_i = \delta_1 + \delta_2 \tag{3. 5}$$

and

$$\sigma_i{}^2 = \sigma_1^2 + \sigma_2^2. \tag{3. 7}$$

Replacing subscript 2 by subscript 3, we also have, for Q_j,

$$\Delta_j = \delta_1 + \delta_3 \tag{3. 11}$$

and

$$\sigma_j{}^2 = \sigma_1^2 + \sigma_3^2. \tag{3. 12}$$

Multiplying equation (3. 11) by (3. 5),

$$\Delta_i\Delta_j = (\delta_1 + \delta_2)(\delta_1 + \delta_3) \tag{3. 13}$$

so that, recalling once more that the X's are uncorrelated,

$$\sigma_{ij} = \sigma_1^2 \tag{3. 14}$$

and

$$\rho_{ij} = \frac{\sigma_1^2}{\sigma_i\sigma_j} = \left(\frac{\sigma_1}{\sigma_i}\right)\left(\frac{\sigma_1}{\sigma_j}\right). \tag{3. 15}$$

Thus, if X_1, X_2, and X_3 are uncorrelated, the common-element correlation is simply the product of two part-whole correlations, and this is the null value of ρ_{ij} under our model.

Exercise

3.1 If ρ_{12}, ρ_{13}, and $\rho_{23} \neq 0$, show that the part-whole correlation is

$$\rho_{ij} = \frac{\sigma_1 + \sigma_{12}}{\sqrt{\sigma_1^2 + \sigma_2^2 + 2\sigma_{12}}},$$

and that the common-element correlation is

$$\rho_{ij} = \frac{\sigma_1^2 + \sigma_{12} + \sigma_{13} + \sigma_{23}}{\sqrt{(\sigma_1^2 + \sigma_2^2 + 2\sigma_{12})(\sigma_1^2 + \sigma_3^2 + 2\sigma_{13})}}.$$

Note that the relation between these results is not nearly as simple as that between equations (3.10) and (3.15). As previously mentioned, retention of covariances in computations of this kind invariably complicates the algebra. In petrology we often seem to forget that this increase in algebraic complexity reflects a similar increase in the complexity of the substantive hypothesis which gives rise to it. Hypotheses which seem reasonable and even forthright when we state them in words may prove utterly unmanageable when we attempt to translate the verbal statement into an algebraic one.

3. Correlation between the Whole and the Ratio of a Part to the Whole [$Q_i = X_1 + X_2, Q_j = X_1/(X_1 + X_2)$]

For Q_i, from the preceding section,

$$\Delta_i = \delta_1 + \delta_2, \tag{3.5}$$

and

$$\sigma_i{}^2 = \sigma_1^2 + \sigma_2^2. \tag{3.7}$$

For the ratio of the part to the whole,

$$Q_j = \frac{X_1}{X_1 + X_2} = \frac{\mu_1 + \delta_1}{(\mu_1 + \mu_2) + (\delta_1 + \delta_2)} \cong \frac{\mu_1 + \delta_1}{\mu_1 + \mu_2}\left[1 - \frac{\delta_1 + \delta_2}{\mu_1 + \mu_2}\right]$$

$$\cong \frac{\mu_1}{\mu_1 + \mu_2} + \frac{\delta_1}{\mu_1 + \mu_2} - \frac{\mu_1}{(\mu_1 + \mu_2)^2}(\delta_1 + \delta_2), \tag{3.16}$$

so that

$$E(Q_j) \cong \frac{\mu_1}{\mu_1 + \mu_2}, \tag{3.17}$$

and

$$\Delta_j = Q_j - E(Q_j) \cong \frac{1}{(\mu_1 + \mu_2)^2} [(\mu_1 + \mu_2)\delta_1 - \mu_1(\delta_1 + \delta_2)]$$

$$\cong \frac{1}{(\mu_1 + \mu_2)^2} (\mu_2\delta_1 - \mu_1\delta_2). \tag{3.18}$$

Squaring equation (3. 18),

$$\Delta_j{}^2 \cong \frac{1}{(\mu_1 + \mu_2)^4} (\mu_2^2\delta_1^2 + \mu_1^2\delta_2^2 - 2\mu_1\mu_2\delta_1\delta_2), \tag{3.19}$$

and, taking expectations on both sides of equation (3. 19),

$$\sigma_j{}^2 = E(\Delta_j{}^2) \cong \frac{1}{(\mu_1 + \mu_2)^4} (\mu_2^2\sigma_1^2 + \mu_1^2\sigma_2^2). \tag{3.20}$$

The product of deviations Δ_i and Δ_j is

$$\Delta_i\Delta_j \cong \frac{1}{(\mu_1 + \mu_2)^2} (\mu_2\delta_1^2 - \mu_1\delta_1\delta_2 + \mu_2\delta_1\delta_2 - \mu_1\delta_2^2), \tag{3.21}$$

so that

$$\sigma_{ij} \cong \frac{1}{(\mu_1 + \mu_2)^2} (\mu_2\sigma_1^2 - \mu_1\sigma_2^2). \tag{3.22}$$

Finally, the ratio of the expected covariance to the geometric mean of the expected variances is

$$\rho_{ij} = \frac{\sigma_{ij}}{\sigma_i\sigma_j} \cong \frac{\mu_2\sigma_1^2 - \mu_1\sigma_2^2}{\sqrt{(\sigma_1^2 + \sigma_2^2)(\mu_2^2\sigma_1^2 + \mu_1^2\sigma_2^2)}}. \tag{3.23}$$

We are here very close to a result of considerable petrographic interest. In petrographic work, however, we would nearly always be dealing with proportions; Q_i would be the sum of two proportions and Q_j the ratio of one of the two to the sum. The null value of ρ_{ij} for this situation is derived in the next section.

4. Correlation between the Whole and the Ratio of a Part to the Whole When Both Are Proportions [$Q_i = Y_1 + Y_2, Q_j = Y_1/(Y_1 + Y_2)$]

Our principal business will be with proportions and, as already announced, the symbol Y is reserved throughout to designate them. Specifically,

$$Y_k = X_k/T, \qquad m > 2,$$

where, for convenience, we set $T = \sum_1^m (X_j)$.

We note first that this requires no change in the approximated mean and variance of Q_j, given in the preceding section, for it is evident that

$$Q_j = \frac{Y_1}{Y_1 + Y_2} = \frac{X_1/T}{(X_1 + X_2)/T} = \frac{X_1}{X_1 + X_2}. \tag{3.24}$$

Hence, as already shown,

$$\Delta_j \cong \frac{1}{(\mu_1 + \mu_2)^2} [\mu_2\delta_1 - \mu_1\delta_2], \tag{3.18}$$

and

$$\sigma_j{}^2 \cong \frac{1}{(\mu_1 + \mu_2)^4} [\mu_2^2\sigma_1^2 + \mu_1^2\sigma_2^2]. \tag{3.20}$$

New approximations are required, however, both for the parameters of Q_i and for its covariance with Q_j.

Denoting the mean of T by τ, and the quantity $(T - \tau)$ by δ_t, we have

$$Y_i = \frac{X_1 + X_2}{T} = \frac{\mu_1 + \mu_2 + \delta_1 + \delta_2}{\tau + \delta_t} \cong \frac{\mu_1 + \mu_2 + \delta_1 + \delta_2}{\tau}\left(1 - \frac{\delta_t}{\tau}\right)$$

$$\cong \frac{\mu_1 + \mu_2}{\tau} + \frac{1}{\tau^2}[\tau(\delta_1 + \delta_2) - (\mu_1 + \mu_2)\delta_t], \tag{3.25}$$

where, as usual, we continue the linear approximation (i.e., in approximating the mean we ignore terms in $\delta_1\delta_t, \delta_2\delta_t$).

Then,

$$E(Y_i) \cong \frac{\mu_1 + \mu_2}{\tau}, \tag{3.26}$$

and

$$\Delta_i \cong \frac{1}{\tau^2}\left[\tau(\delta_1 + \delta_2) - (\mu_1 + \mu_2)\delta_t\right]. \tag{3.27}$$

Squaring equation (3. 27),

$$\Delta_i^2 \cong \frac{1}{\tau^4}\left[\tau^2(\delta_1 + \delta_2)^2 + (\mu_1 + \mu_2)^2\delta_t^2 - 2\tau(\mu_1 + \mu_2)(\delta_1\delta_t + \delta_2\delta_t)\right]. \tag{3.28}$$

Taking expectations on both sides of equation (3. 28), noting that $\delta_t = \delta_1 + \delta_2 + \ldots + \delta_m$, and recalling that, in our null model, $\sigma_{km} = 0$ for all $k \neq m$, we obtain

$$\begin{aligned}\sigma_i^2 &\cong \frac{1}{\tau^4}\left[\{\tau^2 - 2\tau(\mu_1 + \mu_2)\}(\sigma_1^2 + \sigma_2^2) + (\mu_1 + \mu_2)^2\sigma_t^2\right] \\ &\cong \frac{1}{\tau^2}\left[(1 - 2p_1 - 2p_2)(\sigma_1^2 + \sigma_2^2) + (p_1 + p_2)^2\sigma_t^2\right],\end{aligned} \tag{3.29}$$

where $p_k = \mu_k/\tau$, the proportion of $\mathbf{X}_k$ in $\mathbf{X}$ (or probability that a randomly chosen element of $\mathbf{X}$ is of class k).

The product of equation (3. 18) by (3. 27) is

$$\begin{aligned}\Delta_i\Delta_j &\cong \frac{1}{\tau^2(\mu_1 + \mu_2)^2}(\mu_2\delta_1 - \mu_1\delta_2)\left[\tau(\delta_1 + \delta_2) - (\mu_1 + \mu_2)\delta_t\right], \\ &\cong \frac{1}{\tau^2(\mu_1 + \mu_2)^2}\left[\mu_2\tau(\delta_1^2 + \delta_1\delta_2) - \mu_2(\mu_1 + \mu_2)\delta_1\delta_t - \mu_1\tau(\delta_1\delta_2 + \delta_2^2) + \mu_1(\mu_1 + \mu_2)\delta_2\delta_t\right].\end{aligned} \tag{3.30}$$

Taking expectations on both sides of equation (3. 30), recalling once more the zero covariance property of our null model, and replacing μ_k/τ by p_k for $k = 1, 2$,

$$\sigma_{ij} \cong \frac{1 - p_1 - p_2}{(\mu_1 + \mu_2)^2}(p_2\sigma_1^2 - p_1\sigma_2^2). \tag{3.31}$$

The desired null correlation is then

$$\rho_{ij} = \frac{\sigma_{ij}}{\sigma_i\sigma_j} \cong \frac{(1 - p_1 - p_2)[p_2\sigma_1^2 - p_1\sigma_2^2]}{\sqrt{\{p_2^2\sigma_1^2 + p_1^2\sigma_2^2\}\{(1 - 2p_1 - 2p_2)(\sigma_1^2 + \sigma_2^2) + (p_1 + p_2)^2\sigma_t^2\}}}, \tag{3.32}$$

where σ_{ij} is as shown in equation (3. 31), and σ_i, σ_j are, respectively, the square roots of equations (3. 29) and (3. 20).

The numerator of equation (3. 32) will be negative unless $p_2\sigma_1^2 \geqslant p_1\sigma_2^2$. Multiplying both sides of this inequality by τ gives $\mu_2\sigma_1^2 \geqslant \mu_1\sigma_2^2$, and from inspection it is clear that equation (3. 23) will be negative unless this second condition is satisfied. The two null correlations will thus always have the same sign, but in general they will differ in absolute value.

If $X_1 = Ni, X_2 = Fe$, then Q_i and Q_j as defined in the present section (but not in the preceding one!) are the variables of the well known Flight relation, the correlation between the amount and nickel content of the metal phase in stony meteorites (see, for instance, Chayes 1949). If we had some reasonable way to assign values to the p's and σ's, ρ_{ij} from equation (3. 32) would be an appropriate null value against which to test the observed correlation of the Flight variables. A sample value significantly different from that obtained by substitution in equation (3. 32) would overthrow the null hypothesis that the correlation between Ni/(Ni + Fe) and (Ni + Fe) has been generated by the operations of (a) forming percentages out of parent uncorrelated variables and (b) forming a ratio out of these percentages.

We still have much ground to cover before any such test becomes feasible, but it is already clear that $\rho = 0$, the conventional null value, does not provide a meaningful test of the strength of this association, since it can occur, in either equations (3. 23) or (3. 32), if and only if means are proportional to variances in **X**. Up to the linear approximation, at least, the hypothesis that $\rho = 0$ for the Flight correlation is essentially a hypothesis about the relation between means and variances. This is not at all what we have in mind when we plot the Flight variables, or other similar variables, in scatter diagrams.

Exercises

3. 2 If X_2 is a constant, that is, $\sigma_2^2 = 0$, show that:

a. ρ_{ij} from equation (3. 23) is +1. (In interpreting this result, remember that we are using linear approximations!)

b. ρ_{ij} from equation (3. 32) is

$$0 \leqslant \left(\frac{(1 - p_1 - p_2)^2\sigma_1^2}{(1 - p_1 - p_2)^2\sigma_1^2 + (p_1 + p_2)^2(\sigma_t{}^2 - \sigma_1^2)} \right)^{1/2} < 1.$$

3. 3. Denoting the result of equation (3. 23) by ρ_a and that from equation (3. 32) by ρ_b, show that $(\rho_b{}^2/\rho_a^2) \leqslant 1$. Hint: Divide ρ_a^2 by

τ^2/τ^2 and rearrange, to obtain

$$\frac{1}{\sigma_1^2 + \sigma_2^2} \times \frac{(p_2\sigma_1^2 - p_1\sigma_2^2)^2}{p_2^2\sigma_1^2 + p_1^2\sigma_2^2},$$

then factor and rearrange $\rho_b{}^2$ to put it in the form

$$\frac{(1 - p_1 - p_2)^2}{p_2\sigma_1^2 + p_1^2\sigma_2^2}$$

$$\times \frac{(p_2\sigma_1^2 - p_1\sigma_2^2)^2}{(1 - p_1 - p_2)^2(\sigma_1^2 + \sigma_2^2) + (p_1 + p_2)^2(\sigma_t{}^2 - \sigma_1^2 - \sigma_2^2)}$$

Finally, divide the second result by the first.

Note that the closed upper bound on the ratio is required if and only if $(\sigma_t{}^2 - \sigma_1^2 - \sigma_2^2) = 0$, that is, if there are only two variable elements in **X**. Otherwise $\rho_b{}^2$ is always less than ρ_a^2. Show that results (a) and (b) of exercise 3. 2 are compatible with this rule.

3. 4. Using the variables Y, P, T, τ, and $\sigma_t{}^2$ as defined in section 4, show that if $Q_i = Y_1/Y_2$ and $Q_j = Y_1$, then, according to our null model,

(a) $\text{Var}\ (Q_i) \cong \frac{1}{\mu_2^4}(\mu_2^2\sigma_1^2 + \mu_1^2\sigma_2^2)$

(b) $\text{Var}\ (Q_j) \cong \frac{1}{\tau^2}[p_1^2\sigma_t{}^2 + (1 - 2p_1)\sigma_1^2]$

(c) $\text{Cov}\ (Q_i, Q_j) \cong \frac{1}{\mu_2^2}[p_2(1 - p_1)\sigma_1^2 + p_1^2\sigma_2^2]$

and

(d) $\rho_{ij} \cong \frac{p_2(1 - p_1)\sigma_1^2 + p_1^2\sigma_2^2}{\sqrt{\{p_2^2\sigma_1^2 + p_1^2\sigma_2^2\}\{p_1^2\sigma_t{}^2 + (1 - 2p_1)\sigma_1^2\}}}$

Compare equation (d) with equation (2. 8), which is appropriate when the Q's are formed from variables that are not proportions. Note that equation (d) cannot be evaluated unless a sample estimate of $\sigma_t{}^2$ is available. The estimation of $\sigma_t{}^2$ from sample means and variances is described in section 1 of chapter 5, but the method requires a full chemical analysis for each specimen in the sample; in studies of relations between trace and minor elements, therefore, it will often be impossible to use equation (d). In such instances the investigator will be obliged to rely on equation (2. 8) or refrain from attempting interpretation of correlation between a ratio and its numerator or denominator.

5. Correlation between Two Ratios When the Denominator of Each Is the Sum of Its Numerator and a Common Term [$Q_i = X_1/(X_1 + X_2), Q_j = X_3/(X_2 + X_3)$]

This ratio correlation is of particular importance in petrography, for Q_i and Q_j may be regarded as the coordinates of a point in the closed array (Y_1, Y_2, Y_3), where $Y_k = X_k/\Sigma X$; ρ_{ij} will thus be an appropriate null value against which to test the significance of associations shown in ternary diagrams. Before beginning the discussion, let us remember that Q_i and Q_j will have the same values whether defined in terms of Y or X, for,

$$Q_i = \frac{Y_1}{Y_1 + Y_2} = \frac{X_1/\Sigma X}{(X_1 + X_2)/\Sigma X} = \frac{X_1}{X_1 + X_2}, \tag{3.33}$$

and similarly for Q_j.

We have from previous work that, to first-order approximation,

$$E(Q_i) = \frac{\mu_1}{\mu_1 + \mu_2}, \tag{3.17a}$$

$$\delta_i \cong \frac{1}{(\mu_1 + \mu_2)^2} (\mu_2\delta_1 - \mu_1\delta_2), \tag{3.18a}$$

and

$$\sigma_i{}^2 \cong \frac{1}{(\mu_1 + \mu_2)^4} (\mu_2^2\sigma_1^2 + \mu_1^2\sigma_2^2). \tag{3.20a}$$

Similarly,

$$E(Q_j) \cong \frac{\mu_3}{\mu_2 + \mu_3}, \tag{3.34}$$

$$\delta_j \cong \frac{1}{(\mu_2 + \mu_3)^2} (\mu_2\delta_3 - \mu_3\delta_2), \tag{3.35}$$

and

$$\sigma_j{}^2 \cong \frac{1}{(\mu_2 + \mu_3)^4} (\mu_2^2\sigma_3^2 + \mu_3^2\sigma_2^2). \tag{3.36}$$

The product of equation (3.18a) by (3.35) is

$$\delta_i\delta_j \cong \frac{\mu_2^2\delta_1\delta_3 - \mu_1\mu_2\delta_2\delta_3 - \mu_2\mu_3\delta_1\delta_2 + \mu_1\mu_3\delta_2^2}{(\mu_1 + \mu_2)^2(\mu_2 + \mu_3)^2}, \tag{3.37}$$

and, taking expectations on this product, we obtain

$$\sigma_{ij} \cong \frac{\mu_1\mu_3\sigma_2^2}{(\mu_1 + \mu_2)^2(\mu_2 + \mu_3)^2}. \tag{3.38}$$

Then, dividing equation (3.38) by the geometric mean of equations (3.20a) and (3.36),

$$\rho_{ij} \cong \frac{\mu_1\mu_3\sigma_2^2}{\sqrt{(\mu_1^2\sigma_2^2 + \mu_2^2\sigma_1^2)(\mu_2^2\sigma_3^2 + \mu_3^2\sigma_2^2)}}, \tag{3.39}$$

which is identical with line 3 of table 2.1. To the limit of approximation, the term common to both denominators controls the correlation, and this is reasonable enough, since for any pair of values of X_1 and X_3, Q_i and Q_j vary inversely with X_2, whereas variation in X_1 (or X_3) produces no change in Q_j (or Q_i).

As part of the same experiment described in table 2.2 and accompanying discussion, correlations were computed for $Q_i = X_1/(X_1 + X_2)$ and $Q_j = X_3/(X_2 + X_3)$, with the results shown in table 3.1. Evidently the quality of this approximation is much less sensitive to increase in the coefficient of variation than is that of the correla-

Table 3.1 Sample Correlations Between $Q_i = X_1/(X_1 + X_2)$ and $Q_j = X_3/(X_2 + X_3)$ for Different Values of C in an **X** Homogeneous in Mean and Variance (N = 2,000)

C_X	r_{ij}
0.05	0.5023
0.10	0.4847
0.15	0.4907
0.20	0.5006
0.25	0.5046

tions treated in chapter 2. This is fortunate because the general closure test discussed at length in later chapters is inapplicable to the ternary array. The variables of any ternary array may always be converted to those of this section, however, and tested against ρ_{ij} obtained from equation (3. 39).

Exercise

3. 5. Show that according to our null model the appropriate value against which to test the sample correlations of table 3. 1 is $\rho = 0.5$.

4 Correlations between Proportions: The Closed Array

1. The Closure Effect

Suppose that in each of n items we record or measure values for each of m variables, denoting the value of the jth variable in the kth item by Y_{kj}, the average value of the jth variable by $\bar{y}_j$, and the deviation of the jth variable in the kth item by $\Delta_{kj} = Y_{kj} - \bar{y}_j$. Then, regardless of the nature of the variables, we have of course that

$$\sum_{k=1}^{n} \Delta_{kj} = \sum_{k=1}^{n} (Y_{kj} - \bar{y}_j) = \sum_{k=1}^{n} Y_{kj} - n\bar{y}_j = 0,$$

for, by definition, $\bar{y}_j = \frac{1}{n} \sum_{k=1}^{n} Y_{kj}$.

The observations may be thought of as forming an $n \times m$ matrix in which each row vector $\mathbf{Y}_k$, contains m entries, one for each variable observed in the kth item. Replacing each Y_{kj} by its appropriate Δ_{kj}, we then have that the column sums of the (deviation) matrix are all zero.

With rows and columns as defined here, the row sums of the deviation matrix are ordinarily of no interest; usually they are not even explicitly defined. If, however, the data are such that the sum of all variables in each item is a constant, the same for all items in the sample—that is, $\sum_{j=1}^{m} Y_{kj} = K$ for all k—then, since

$$\Delta_{kj} = Y_{kj} - \bar{y}_j = Y_{kj} - \frac{1}{n} \sum_{k=1}^{n} Y_{kj},$$

we have, summing both sides over j, that

$$\sum_{j=1}^{m} \Delta_{kj} = \sum_{j=1}^{m} Y_{kj} - \frac{1}{n} \sum_{k=1}^{n} \sum_{j=1}^{m} Y_{kj} = K - \frac{1}{n} \sum_{k=1}^{n} K = 0,$$

the order of summation in the double sum being of no consequence. Thus the row sums of the deviation matrix are also zero; the "closed array" has the curious property that

$$\sum_{k=1}^{n} \Delta_{kj} = \sum_{j=1}^{m} \Delta_{kj} = 0. \tag{4.1}$$

Omitting the row subscript, we may obtain from any row of this matrix the equation

$$\Delta_1 + \Delta_2 + \cdots + \Delta_i + \cdots + \Delta_m = 0.$$

Multiplying through by any individual deviation, say Δ_i, and taking expectations, we have immediately that

$$\text{var}\ (Y_i) + \sum_{j \neq i} \text{cov}\ (Y_i, Y_j) = 0, \qquad j \neq i, \tag{4.2}$$

so that the sum of the covariances of any variable is negative. Thus each variable must be negatively correlated with at least one other variable; and, in general, there is a strong bias toward negative correlation between variables of (relatively) large variance. Since prior to any computation—indeed, prior even to any inspection—of a closed array, we may be quite sure that the net covariance of every variable will be negative, tests of the strength of association against null values calculated without regard for the effect of closure on correlation, and in particular the conventional test against $\rho = 0$, are clearly inappropriate. In this chapter we derive approximations of null correlations which allow for the effect of closure on the initially unconstrained and uncorrelated variables of the null model proposed in chapter 1. The key relation is an obvious extension of Pearson's original definition of spurious correlation, namely the correlation between ratios with common denominator, for which see line 3 of table 2.1.

2. Null Correlation between Ratios Whose Numerators Are Common Elements in Their Common Denominator ($Y_i = X_i/T, Y_j = X_j/T, T = \sum_1^m X$)

For variables defined in this way it is evident that

$$Y_i = \frac{X_i}{T} = \frac{\mu_i + \delta_i}{\tau + \delta_t} \cong \frac{\mu_i + \delta_i}{\tau}\left(1 - \frac{\delta_t}{\tau}\right) \cong p_i + \frac{1}{\tau}(\delta_i - p_i\delta_t), \tag{4.3}$$

where $p_i = \mu_i/\tau$, the proportion or probability of X_i in **X**.

Hence

$$E(Y_i) \cong p_i. \tag{4.4}$$

In order to avoid confusion arising from the fact that each (X, Y) pair is now identified by a common subscript, we extend the dual notation adopted in the preceding section in the following way:

	X	Y
Deviations from mean	δ_i, δ_j	Δ_i, Δ_j
Variances	σ_i^2, σ_j^2	var (Y_i), var (Y_j)
Covariances	σ_{ij}	cov (Y_i, Y_j)

Subtracting equation (4. 4) from (4. 3), we have in this notation that

$$\Delta_i \cong \frac{1}{\tau}(\delta_i - p_i\delta_t). \tag{4.5}$$

The square of equation (4. 5) is

$$\Delta_i^2 \cong \frac{1}{\tau^2}(\delta_i^2 - 2p_i\delta_i\delta_t + p_i^2\delta_t^2),$$

and, taking expectations in the usual fashion,

$$\text{var}\ (Y_i) \cong \frac{1}{\tau^2}[p_i{}^2\sigma t^2 + (1 - 2p_i)\sigma_i^2]. \tag{4.6}$$

Similarly, replacing i by j in equations (4. 5) and (4. 6),

$$\Delta_j \cong \frac{1}{\tau}(\delta_j - p_j\delta_t), \tag{4.7}$$

and

$$\text{var}\ (Y_j) \cong \frac{1}{\tau^2}[p_j^2\sigma_t^2 + (1 - 2p_j)\sigma_j^2]. \tag{4.8}$$

The product of equation (4. 5) by (4. 7) is

$$\Delta_i\Delta_j \cong \frac{1}{\tau^2}(\delta_i\delta_j - p_j\delta_i\delta_t - p_i\delta_j\delta_t + p_ip_j\delta_t^2),$$

so that, again taking expectations and rearranging,

$$\text{cov}\ (Y_i, Y_j) \cong \frac{1}{\tau^2}(p_ip_j\sigma_t^2 - p_j\sigma_i^2 - p_i\sigma_j^2). \tag{4.9}$$

Finally, the parent correlation according to our null model, and referred to subsequently as "closure" correlation, is

$$\rho_{ij} = \frac{\text{cov}\,(Y_i, Y_j)}{\sqrt{\text{var}\,(Y_i) \times \text{var}\,(Y_j)}}$$

$$\cong \frac{p_i p_j \sigma_t^2 - p_j \sigma_i^2 - p_i \sigma_j^2}{\sqrt{[p_i^2 \sigma_t^2 + (1 - 2p_i)\sigma_i^2]\,[p_j^2 \sigma t^2 + (1 - 2p_j)\sigma_j^2]}}$$

(4. 10)

A sample correlation differing significantly from ρ_{ij} would overthrow the null hypothesis, establishing, in a probability sense, that the observed correlation could not result merely from the closure of originally uncorrelated open variables. As is usual in all such tests, of course, a sample correlation which does not differ significantly from ρ_{ij} does not, in the sense of ordinary language, confirm the null hypothesis; it is merely compatible with, or provides no evidence against, that hypothesis.

We have so far ignored the ugly fact that before any test is possible it will be necessary to assign reasonable values to the parameters of **X**, the p's and σ's which appear on the right side of equation (4. 10). The reader will recall that the same difficulty arose, and was also ignored, in connection with equation (3. 32) of chapter 3. It will be treated in detail in chapter 5; the remainder of the present chapter is concerned with some of the more important properties of equation (4. 10), properties whose theoretical interest is evident even though their practical application must be deferred pending the work of chapter 5.

3. The Sign of the Closure Correlation

We have already seen that there is a strong bias toward negative correlation in the closed array, that at least one of the correlations of each variable in such an array must be negative, and that the net covariance of every variable is negative. It does not follow, however, that the closure correlation, the correlation imposed upon any two variables of an uncorrelated open array by the operation of closure, is necessarily negative. The sign of the closure correlation is governed by the sign of the numerator of equation (4. 10), namely, $p_i p_j \sigma_t^2 - p_j \sigma_i^2 - p_i \sigma_j^2$. All quantities in this expression being by definition positive, ρ_{ij} will be positive or negative depending on whether $p_i p_j \sigma_t^2$ is greater or less than $(p_j \sigma_i^2 + p_i \sigma_j^2)$. Since only one of the p's may exceed 0. 5, the product of any two will usually be much smaller than either one separately, and ρ_{ij} will not be positive unless the sum of the other variances is much larger than $(\sigma_i^2 + \sigma_j^2)$. It is thus likely that positive closure correlation will be confined to variables whose open variances are, relatively, very small.

Exercises

4.1 From inspection of equation (4.8) notice that var $(Y_j) > 0$ even if $\sigma_j^2 = 0$, that is, Y_j is a variable even though X_j is a constant.

4.2 By substitution in equation (4.10) show that if $\sigma_i^2 = \sigma_j^2 = 0$, $\rho_{ij} = +1$.

4.3 By substitution in the numerator of equation (4.10) show that:
  a. if means are homogeneous in **X**, that is, if $p_1 = p_2 = \ldots = p_m = 1/m$, ρ_{ij} will be positive if $(\sigma_i^2 + \sigma_j^2) < \sigma_t^2/m$, the average variance of the array;
  b. if variances are homogeneous in **X**, ρ_{ij} will be positive if and only if $m > (p_i + p_j)/p_i p_j$;
  c. if both means and variances are homogeneous in **X** no closure correlation will be positive.

4. Closure Correlation When Variances and Means are Homogeneous in X ($\sigma_i^2 = \omega$, $\mu_i = \theta$, for all i)

This is an important special case in which the closure correlation takes a particularly simple form. If $\mu_i = \theta$ for all i, then, of course,

$$p_i = \mu_i/\tau = \theta/m\theta = 1/m. \tag{4.11}$$

Further, ignoring a common scaling factor, we have from equation (4.8) that

$$\text{var}\ (Y_j) = \text{var}\ (Y_i) \cong \left(1 - \frac{1}{m}\right)\omega,$$

and from equation (4.9) that (4.12)

$$\text{cov}\ (Y_i, Y_j) = -\omega/m.$$

Thus,

$$\rho_{ij} \cong \frac{\text{cov}\ (Y_i, Y_j)}{\sqrt{\text{var}\ (Y_i) \times \text{var}\ (Y_j)}} = -\frac{1}{m-1}. \tag{4.13}$$

This result may of course also be obtained directly from equation (4.10). The specification of homogeneity of means is mandatory, for if the means are not homogeneous in **X**, closure of **X** will not generate a Y whose variances are homogeneous. It is clear from equation

(4. 4), furthermore, that if the means are indeed homogeneous in **X** they will also be homogeneous in **Y**.

Exercise

4. 4 Show that if variances are homogeneous but means are not homogeneous in **X**, neither variances nor means will be homogeneous in **Y**. [Heterogeneity of means of **Y** follows at once from equation (4. 4) and of variances from (4. 6).]

5. Closure Correlation When Variances Are Proportional to Means in X ($\sigma_i^2/\mu_i = \kappa$ for all i)

A parent distribution in which variances are proportional to means is of considerable interest, both theoretical and practical. Mosimann has recently shown that if this relation holds in **X**, then $\rho_{ij} = -[p_i p_j/(1-p_i)(1-p_j)]^{1/2}$ exactly. It is comforting to discover that this same result can also be obtained from our first-order approximation.

If $\sigma_i^2/\mu_i = \kappa$, then, multiplying both sides of the condition by τ,

$$\frac{\tau\sigma_i^2}{\mu_i} = \frac{\sigma_i^2}{\mu_i/\tau} = \frac{\sigma_i^2}{p_i} = \tau\kappa, \qquad 1 \leqslant i \leqslant m.$$

Further, since $\sigma_t^2 = \Sigma\sigma_i^2$ and $p_t = \Sigma p_i = 1$, we also have $\sigma_t^2 = \tau\kappa$. Substituting these results into equation (4. 10),

$$\rho_{ij} \cong \frac{-\kappa\tau p_i p_j}{\sqrt{\kappa^2\tau^2 p_i p_j (1-p_i)(1-p_j)}} = -\sqrt{\frac{p_i p_j}{(1-p_i)(1-p_j)}}, \tag{4. 14}$$

which is Mosimann's result.

Exercises

4. 5 Mosimann also shows that if variances are proportional to means in **X**, the variances of **Y** will be proportional to $p(1-p)$ and cov (Y_1, Y_j) will be proportional to $-p_i p_j$. Obtain these results by substitution in equations (4. 6) and (4. 9), respectively.

4. 6 Note that exercise 4. 4 is a special case of the reduction leading to equation (4. 14), with $\kappa = m\theta$.

6. The Ternary Closed Array

Although of immense practical concern to petrographers, the closed array with $m = 3$ is of interest here chiefly as a further test of equation (4.10), for the "closure" correlations in any ternary closed array can be calculated exactly from the (closed) variances alone and turn out to be identical with the approximations obtained from equation (4.10).

If, in **Y**, m = 3, we may write any row of the deviation matrix as

$$\Delta_1 + \Delta_2 = -\Delta_3.$$

Squaring both sides and rearranging terms,

$$\Delta_1\Delta_2 = \frac{1}{2}\,(\Delta_3^2 - \Delta_1^2 - \Delta_2^2). \tag{4.15}$$

Taking expectations, we have at once that

$$\mathrm{cov}\,(Y_1, Y_2) = \frac{1}{2}\,[\mathrm{var}\,(Y_3) - \mathrm{var}\,(Y_1) - \mathrm{var}\,(Y_2)], \tag{4.16}$$

and, dividing by the geometric mean of the variances,

$$\rho_{12} = \frac{1}{2}\,\frac{\mathrm{var}\,(Y_3) - \mathrm{var}\,(Y_1) - \mathrm{var}\,(Y_2)}{\sqrt{\mathrm{var}\,(Y_1) \times \mathrm{var}\,(Y_2)}} \tag{4.17}$$

in which, of course, subscripts may be rotated.[1]

To show that equation (4.17) is equivalent to (4.10) it is necessary to prove that, if m = 3, the right side of equation (4.16) is indeed equal to the bracketed term on the right side of (4.9). I outline the proof here, leaving the detailed algebra as an exercise for the interested reader. Given m = 3 and $\sigma_{ij} = 0$ for $i \neq j$, we have that $p_3 = 1 - p_1 - p_2$ and $\sigma_t^2 = \sigma_1^2 + \sigma_2^2 + \sigma_3^2$. Then, from equation (4.6), ignoring the proportionality constant,

$$\begin{aligned}
\mathrm{var}\,(Y_1) &= p_1^2\sigma_2^2 + p_1^2\sigma_3^2 + (1 - p_1)^2\sigma_1^2,\\
\mathrm{var}\,(Y_2) &= p_2^2\sigma_1^2 + p_2^2\sigma_3^2 + (1 - p_2)^2\sigma_2^2,\\
\mathrm{var}\,(Y_3) &= p_3^2\sigma_1^2 + p_3^2\sigma_2^2 + (1 - p_3)^2\sigma_3^2.
\end{aligned} \tag{4.18}$$

---

[1] This formulation is also applicable to the computation of sample correlations, except that instead of taking expectations to get from equation (4.15) to (4.16) we sum the right side of equation (4.15) over the n rows of the sample deviation matrix and divide each cumulated sum by (n − 1), to obtain the quantity $(s_3^2 - s_1^2 - s_2^2)/2$. Then we have at once that $r_{12} = (s_3^2 - s_1^2 - s_2^2)/2s_1s_2$, a result reached without explicit calculation of the covariance. In practice, of course, we would simply compute the three variances directly, and not bother with equation (4.15) at all.

Replacing each p_3 in equation (4.18) by $(1 - p_1 - p_2)$ and subtracting the first two lines from the third, we have, after a rather elaborate drill in elementary algebra, that

$$\frac{1}{2}\,[\mathrm{var}\,(Y_3) - \mathrm{var}\,(Y_1) - \mathrm{var}\,(Y_2)] = p_1 p_2 \sigma_t^2 - p_2 \sigma_1^2 - p_1 \sigma_2^2, \tag{4.19}$$

as required.

This result is of some interest in later work because of the specification that, in **X**, $\sigma_{ij} = 0$ for all $i \neq j$. We are thus always free to transform the variables $(\mu_i, \mu_j, \ldots, \mu_m)$, with variances $(\sigma_i^2, \sigma_j^2, \ldots, \sigma_m^2)$, into

$$\mu_1 = \mu_i,\ \mu_2 = \mu_j, \text{ and } \mu_3 = \sum_{k=1}^{m} \mu_k, \qquad k \neq i \text{ or } j,$$

with variances

$$\sigma_1^2 = \sigma_i^2,\ \sigma_2^2 = \sigma_j^2, \text{ and } \sigma_3^2 = \sum_{k=1}^{m} \sigma_k^2, \qquad k \neq i \text{ or } j.$$

For the purpose of calculating the closure correlation it would thus appear that we could reduce any m-variable array to a ternary one, and use equation (4.17), which is exact, rather than (4.10), which is an approximation. Application of equation (4.17), of course, requires that we know var (Y_i) for $i = 1, 2, 3$; and we do not yet have this information. In the rather restricted context of the test proposed in the next chapter solutions of equation (4.10) do indeed seem to be exact rather than approximate. It will be shown in chapter 6 that this advantage is more apparent than real; equation (4.17) is always an appropriate device for calculating the correlations in samples of any ternary closed array but does not yield meaningful null values. In our null model, the operation of comparing an observed to an expected correlation is meaningful only if $m > 3$. The variables of any ternary array may be subjected to the transformation described in section 5 of chapter 3, however, and then tested against the null correlation obtained by substitution in equation (3.39).

5 Assigning Values to the Parameters of the Hypothetical Open Array

In the work of the preceding chapters we have taken the view that in order to understand and allow for the effects of ratio formation and closure on correlation we must investigate the effects of these operations on initially uncorrelated variables. It is difficult to fault this logic, at least for the purpose of an elementary approach to the subject. But it is time to point out that, before the understanding so achieved can be applied in practical work, we must have a more than symbolic knowledge of the parameters of the underlying array of open, uncorrelated variables. This is a difficult requirement to satisfy, particularly puzzling in connection with the proportions usually encountered in chemical petrography, because we never observe the raw values from which these proportions are formed.

In classical rock analysis, for instance, one may determine the alkalis on a half-gram sample, silica and the oxides of the R_2O_3 group on a separate gram of material, FeO on still a third portion, P_2O_5 on a fourth, and so on. In any meaningful statement of bulk composition these results must be "normalized" to parts of some common weight or volume. For this purpose the dimensions of the sample powdered for analysis, like those of the hand specimen from which it is obtained, are never of more than trivial interest. This is true also of the outcrop from which the specimen is taken and of the exposure area or map unit as well. The sensible standard "weight" is unity, so that the results are simply proportions. For the more complex purpose of comparing two analyses, whether of the same or different specimens or rocks, scaling to unity is unavoidable. Under proper circumstances analyses may be regarded as samples of **Y**; they are never samples of **X**.

Yet correlations between each pair of variables in a series of analyses will be strongly influenced by the closure restraint, and if we wish to allow for this effect in appraising the strength of association we shall have to be able to assign values to the means and variances of **X**, values which are both possible numerically and reasonable substantively, even though no direct observation of **X** is possible; and, physically, **X** may not even exist.

In this chapter we present a method of obtaining what appear to be substantively reasonable values for the parameters of **X**, parameters from which, in the succeeding chapter, we compute null values against which observed correlations may be tested.

1. Relations between the Means and Variances of X and Y

Our principal concern is to avoid attributing substantive significance to correlations which may reflect nothing more than the interdependence of variances and covariances arising from closure. As a first step we construct the simplest possible model which will provide information bearing on this situation. In brief, we find values for the parameters of **X** such that the closure of **X** will yield a **Y** whose means and variances are exactly those of the sample. Since $\sigma_{ij} = 0$, by definition, for all i in **X**, the covariances of **Y** will have been generated entirely by the operation of closure; the correlations in **Y** are thus reasonable null values against which to compare the observed correlations. An observed correlation, r_{ij}, that differs significantly from ρ_{ij}, its analogue in **Y**, overthrows our null hypothesis—that the sample correlation might be generated by closure—and from it we conclude that, at some announced probability level, there is indeed a substantively meaningful, that is, a nonrandom, association between variables i and j; whenever r_{ij} does not differ significantly from ρ_{ij}, we conclude that the sample provides no evidence warranting rejection of the null hypothesis.

The covariances of **X** all being zero, we have only to find its means and variances. The assignment of means offers no difficulty. We have already seen that $E(Y_i) = \mu_i/\tau = p_i$. In words, the mean value of Y_i, to the limit of our approximation, is the mean proportion (or probability) of X_i in **X**. Thus, if we want the mean of Y_i to be the same as that of the sample, we must set p_i in **X** equal to the observed average.

The choice of variances for **X** is considerably more complicated, but the initial formulation is identical. We already have an approximation of var (Y_i) as a function of the means and variances of **X**, and we want it to be equal to the sample variance of variable i, denoted by s_i^2. This requirement is symbolized by setting the right side of equation (4. 6) equal to s_i^2, or

$$\frac{1}{\tau^2}\{p_i^2\sigma_t^2 + (1 - 2p_i)\sigma_i^2\} = s_i^2, \qquad i = 1, m. \qquad (5.1)$$

Solving equation (5.1) explicitly for σ_i^2,

$$\sigma_i^2 = \frac{1}{1 - 2p_i} \{\tau^2 s_i^2 - p_i^2 \sigma_t^2\}, \tag{5.2}$$

and it is clear, if circular, that we have to know σ_t^2 before we can find σ_i^2. Recalling that $\sigma_t^2 = \sum_{i=1}^{m} \sigma_i^2$, however, we sum equation (5.2) over i to obtain

$$\sigma_t^2 = \sum_{i=1}^{m} \left\{\frac{\tau^2 s_i^2 - p_i^2 \sigma_t^2}{1 - 2p_i}\right\}, \qquad p_i \neq 1/2, \tag{5.3}$$

or, rearranging terms to give an explicit solution,

$$\sigma_t^2 = \frac{\tau^2 \sum_{i=1}^{m} \left(\frac{s_i^2}{1 - 2p_i}\right)}{1 + \sum_{i=1}^{m} [p_i^2/(1 - 2p_i)]}, \qquad p_i \neq 1/2 \tag{5.4}$$

This may be further simplified to

$$\sigma_t^2 = \frac{\tau^2 \sum_{i=1}^{m} \left(\frac{s_i^2}{1 - 2p_i}\right)}{\sum_{i=1}^{m} \left(\frac{p_i(1 - p_i)}{1 - 2p_i}\right)}, \qquad p_i \neq 1/2, \tag{5.5}$$

since $\Sigma p_i = 1$. Then, substituting equation (5.5) for σ_t^2 in (5.2),

$$\sigma_i^2 = \frac{1}{1 - 2p_i} \left\{ s_i^2 - p_i^2 \frac{\sum_{j=1}^{m} \left(\frac{s_j^2}{1 - 2p_j}\right)}{\sum_{j=1}^{m} \left(\frac{p_j(1 - p_j)}{1 - 2p_j}\right)} \right\}, \qquad p \neq 1/2, \tag{5.6}$$

where we set $\tau = 1$ to put var (Y_i) and s_i^2 in the same scale, and, to avoid confusion, sum over subscript j instead if i.

This bootstrap operation works, of course, only if the denominator of the term in p_i^2 is other than zero, and we next show that this will always be so if $m > 2$. It is evident that if $p_j < 1/2$ for all j—which

Then, left-multiplying both sides of equation (5.14) by the inverse of **P**, the vector of open variances is

$$\boldsymbol{\sigma} = \mathbf{P}^{-1}\mathbf{s}, \tag{5.15}$$

and the ith open variance is

$$\sigma_i = {}^i\mathbf{P}\mathbf{s}, \tag{5.16}$$

where ${}^i\mathbf{P}$ is the ith row of $\mathbf{P}^{-1}$. The passage from equation (5.14) to (5.15) of course assumes nonsingularity of **P**, but this is already established by the work of the last section, which shows that, in principle, explicit solution for any σ_i^2 is always possible.

From the sample statistics, we are thus able to assign means and variances to **X** such that **Y** formed by closure of **X** will have expected means and variances identical with those observed, *but covariances generated entirely by the operation of closure.*

Exercises

5.3 Readers who enjoy algebra may like to prove nonsingularity of **P** by showing that $|\mathbf{P}| \neq 0$. (Hint: Evaluate $|\mathbf{P}|^2$ separately for m even and odd. For each case show that $|\mathbf{P}|$ has a minimum, and that this minimum is greater than zero.)

5.4 Working with either equation (5.6) or a computer program based upon equation (5.15), use the observed means and variances in the display below to obtain the theoretical open variances shown in the last column.

Variable		**Sample Statistics**		**Open Variances**
Name	Number (i)	$10^2\bar{y}_i$	$10^4 s_i^2$	$10^4\sigma_i^2$
Quartz	1	29.37	19.5021	21.8262
Microcline	2	34.19	35.7878	68.1950
Plagioclase	3	29.85	20.5712	24.1408
Biotite	4	4.50	6.1757	6.5157
Muscovite	5	2.08	1.0189	1.0082

(If you use a desk computer the entries in the last column probably will not check your results to more than 4 figures. The multipliers 10^2 and 10^4 are used in the display both as a convenience and because we usually think of modes as percentages rather than proportions. In performing the calculations, remem-

implies that $m > 2$—every term in this sum, and hence the whole sum, is positive. But if $m = 2$, the sum is simply

$$A_2 = \frac{p_1(1-p_1)}{1-2p_1} + \frac{p_2(1-p_2)}{1-2p_2} = \frac{p_1(1-p_1)}{1-2p_1} - \frac{p_1(1-p_1)}{1-2p_1} = 0, \tag{5.7}$$

and some p, say p_1, must be greater than ½. If we then partition X_2 into X_a and X_b, with mean proportions $a + b = p_2$, its contribution, C_2, to the sum is replaced by

$$C_{a+b} = \frac{a(1-a)}{1-2a} + \frac{b(1-b)}{1-2b} = \frac{(a+b-2ab)(1-a-b)}{1-2a-2b+4ab}, \tag{5.8}$$

where C_{i+j} denotes the contribution of the ith and jth terms, C_i and C_j. But, since $p_2 = a + b$, the contribution from X_2 prior to its partition may be written

$$C_2 = \frac{p_2(1-p_2)}{1-2p_2} = \frac{(a+b)(1-a-b)}{1-2a-2b}, \tag{5.9}$$

and it is clear from inspection that $C_{a+b} < C_2$. It follows then, the sum in the denominator of equation (5.6) being denoted by $A_2 = C_1 + C_2$ if $m = 2$ and $A_3 = C_1 + C_a + C_b$ if $m = 3$, that we shall always have $A_3 < A_2 = 0$, providing only that $p_2 = a + b$. Partitioning one of the new variables, say X_b, into $X_d + X_e$ with $d + e = b$, we have, in exactly the same fashion, that $C_{d+e} < C_b$, from which $A_4 < A_3 < A_2 = 0$, and so forth. Thus, quite generally,

$$A_m = \sum_{j=1}^{m}\left(\frac{p_j(1-p_j)}{1-2p_j}\right) \neq 0, \qquad m > 2,\ p_j \neq ½,$$

A_m being negative or positive depending on whether or not some one of the p's is greater than ½. Accordingly, explicit solution for σ_i^2 is always possible, subject only to the restrictions that $m > 2$, and $p_j \neq ½$ for all j. The first restriction is trivial, since we are not going to be deeply concerned about correlation in a two-variable closed array. The second is scarcely likely to cause much trouble in practical work, but in principle the right side of equation (5.6) is undefined if some $p = ½$. For the sake of completeness we show that explicit solutions for the σ_i^2 may then be obtained directly from equations (5.1) and (5.2). If, say, $p_j = ½$, we have at once from equation (5.1) that $\sigma_t^2 = 4s_j^2$. Substituting this in the right side of equation (5.2) gives

$$\sigma_i^2 = \frac{1}{1 - 2p_i} (s_i^2 - 4p_i^2 s_j^2), \qquad m > 2,\ i \neq j, \qquad (5.10)$$

and, finally, by difference,

$$\sigma_j^2 = 4s_j^2 - \sum_{i=1}^{m} \left(\frac{s_i^2 - 4p_i^2 s_j^2}{1 - 2p_i} \right), \qquad m > 2,\ i \neq j, \qquad (5.11)$$

Accordingly, explicit solutions can always be found for σ_i^2 when $m > 2$. On occasion, however, one or more of these solutions will be negative, so that even though the closure of **X** must always yield the desired relation between the parameters of **Y** and the statistics computed from the sample, **X** itself may be impossible. This puzzling situation is examined further in chapters 6 and 9.

Exercises

5.1 Generate the fourth column of the following display by using the first three as input to equations (5.10) and (5.11):

i	p_i	$10^3 s_i^2$	$10^3 \sigma_1^2$
1	0.5	2.0	3.50
2	0.3	2.5	4.45
3	0.1	0.1	0.025
4	0.1	0.1	0.025

5.2 Generate the fourth column of the following display by using the first three as input to equation (5.6):

i	p_i	$10^3 s_i^2$	$10^3 \sigma_i^2$
1	0.49	2.0	3.4486
2	0.31	2.5	4.5450
3	0.10	0.1	0.0245
4	0.10	0.1	0.0245

Although equation (5.6) cannot be used in exercise 5.1, p_1 of exercise 5.2 is close enough to 0.5 so that equations (5.10) and (5.11) give quite fair results. Compute them.

2. An Alternative Method for Computing the Variances of X

If the number of variables is large the calculations described in preceding section may be quite tedious to perform on an ordinar desk calculator. On electronic desk calculators with as many as two memory cells they are no more than a few minutes work, an for a programmed computer they are, of course, trivial. But if o of the latter is available, it may be more convenient to formulate the calculation in matrix algebra. For this purpose, and because shall require the matrix formulation in chapter 9, we restate equation (5.1) as

$$(1 - p_i)^2 \sigma_i^2 + p_i^2 \sum_{j \neq i}^{m} (\sigma_j^2) = s_i^2, \qquad j \neq i, \qquad (5.12)$$

so that for each i we have an equation of the form

$$p_i^2 \sigma_1^2 + p_i^2 \sigma_2^2 + \ldots + (1 - p_i)^2 \sigma_i^2 + \ldots + p_i^2 \sigma_m^2 = s_i^2, \qquad (5.13)$$

or, in matrix notation,

$$\mathbf{P}\boldsymbol{\sigma} = \mathbf{s}, \qquad (5.14)$$

where

$$\boldsymbol{\sigma} = \{\sigma_1^2, \sigma_2^2, \ldots, \sigma_m^2\},$$

$$\mathbf{s} = \{s_1^2, s_2^2, \ldots, s_m^2\},$$

and **P** is the m × m coefficient matrix

$$\begin{bmatrix} (1 - p_1)^2 & p_1^2 & p_1^2 & \cdots & p_1^2 \\ p_2^2 & (1 - p_2)^2 & p_2^2 & \cdots & p_2^2 \\ p_3^2 & p_3^2 & (1 - p_3)^2 & \cdots & p_3^2 \\ \vdots & \vdots & \vdots & & \vdots \\ p_m^2 & p_m^2 & p_m^2 & & (1 - p_m^2) \end{bmatrix}$$

ber to set $p_i = \bar{y}_i$, not $10^2 \bar{y}_i$. The positioning of the decimal point in s_i^2 is immaterial; σ_i^2 and s_i^2 will be in the same scale.)

The data, which will be used again in chapter 6, are modal analyses of a suite of 15 thin sections of the Bellingham, Minnesota, granite. The display is taken from table 1 of Chayes and Kruskal (1966), where the next to last entry in the last column is mistakenly given as 5.5157.

6 Testing the Significance of an Observed Correlation between Proportions

In chapter 4, the closure correlation, defined as the correlation to be expected between $Y_i = X_i/T$ and $Y_j = X_j/T$, where $T = \Sigma X$ and the covariance between X's, σ_{ij}, is zero for all $i \neq j$, was found to be

$$\rho_{ij} \cong \frac{p_i p_j \sigma_t^2 - p_j \sigma_i^2 - p_i \sigma_j^2}{\sqrt{[p_i^2 \sigma_t^2 + (1 - 2p_i)\sigma_i^2][p_j^2 \sigma_t^2 + (1 - 2p_j)\sigma_j^2]}}. \quad (4.10)$$

No practical use of ρ_{ij} was possible, however, since the p's and σ^2's required to compute it are parameters of **X**, and these were unknown.

In chapter 5 we formulated a procedure for assigning values to the parameters of **X** such that the Y's resulting from closure of **X** would have means and variances equal to those observed in any actual sample. With these values for the mean proportion, p_i, and variance, σ_i^2, of each element of **X**, we now use equation (4.10) to obtain numerical values of ρ_{ij} suitable for testing the null model proposed in chapter 1. The general procedure for testing correlation coefficients is then briefly reviewed and applied to an example of closure correlation. The chapter concludes with some notes about the limitations of the particular test proposed here.

1. Calculating the Closure Correlation

It will be recalled from chapter 5 that the means and variances of the parent **Y** will agree, to the limits of our approximation, with those of the sample if we set p_i in **X** equal to $\bar{y}_i$, the observed mean of Y_i, and

$$\sigma_i^2 = \frac{1}{1 - 2p_i} \left\{ s_i^2 - p_i^2 \frac{\sum_{j=1}^{m} [s_j^2/(1 - 2p_j)]}{\sum_{j=1}^{m} [p_j(1 - p_j)/(1 - 2p_j)]} \right\}, \quad (5.6)$$

for all i. Then, since $\sigma_{ij} = 0$ for all $i \neq j$, we also have that $\sigma_t^2 = \Sigma_i \sigma_i^2$, and if no $\sigma_i^2 < 0$ we may rewrite equation (4.10), the closure correlation, as

$$\rho_{ij} \cong (\bar{y}_i \bar{y}_j \sigma_t^2 - \bar{y}_j \sigma_i^2 - \bar{y}_i \sigma_j^2)/s_i s_j, \tag{6.1}$$

in which $\bar{y}_i$ is the observed average of the ith variable, s_i is the standard deviation of variable i computed from the sample, σ_i^2 is the estimated open variance of variable i in our null model, obtained from equation (5.6), and similarly for $\bar{y}_j$, s_j, and σ_j^2.

Exercise

6.1 Using equation (6.1) and the data display of exercise 5.4, calculate the closure correlation of quartz with microcline ($\rho_{12} \cong -0.5781$) and of microcline with plagioclase ($\rho_{23} \cong -0.5697$).

2. The Fisher z Transformation[1]

For samples of size n drawn randomly from a parent in which X_i and X_j are jointly normally distributed with $\rho = 0$, the quantity

$$r\sqrt{(n-2)/(1-r^2)} \tag{6.2a}$$

is distributed as Student's t with (n − 2) degrees of freedom, and the ratio

$$F' = r^2(n-2)/(1-r^2) = t'^2 \tag{6.2b}$$

is distributed as Snedecor's F with 1 degree of freedom in the numerator and (n − 2) degrees of freedom in the denominator. The null hypothesis, that $\rho = 0$, is rejected or retained, at the α level, depending on whether or not $t' > t(\alpha, n-1)$; the same result will be found if F' is compared with $F^1_{n-1}(\alpha)$.

Unfortunately, the test is applicable only against the alternative that $\rho = 0$, for otherwise the distribution of r is badly skewed even in rather large samples. This book is mostly concerned with situations in which $\rho \neq 0$; although it is then inconvenient to test r directly, R. A. Fisher discovered that, except for $|\rho|$ in the immediate vicinity of $|1|$, the quantity

$$z = \frac{1}{2}[\ln(1+r) - \ln(1-r)] = \tanh^{-1} r \tag{6.3}$$

[1] Readers familiar with correlation analysis should pass directly to the next section.

is very nearly normally distributed with variance $(n - 3)^{-1}$. To decide whether a sample correlation, r, warrants rejection, at the α level, of the null hypothesis that the parent correlation is some $\rho \neq 0$, we compare the statistic

$$t'' = |\tanh^{-1} r - \tanh^{-1} \rho| \sqrt{n - 3} \qquad (6.4)$$

with t_α drawn from a table of the cumulative normal frequency distribution (or, alternatively, with Student's t_α for infinite degrees of freedom). The null hypothesis that the parent correlation is ρ is rejected or retained depending on whether or not $t'' > t_\alpha$.

(This résumé of well-known testing procedures is inserted to reduce the necessity for extensive outside reference on the part of those insufficiently familiar with elementary statistics. The uninitiated reader who plans to use the test of closure correlation suggested here in his own work should certainly first study the chapter(s) on correlation in a standard text.)

Fisher's derivation of the z test presumes a bivariate normal parent, but it has since been found that the statistic z is "robust" in the sense that its distribution may be little affected by rather considerable departures from normality in the X's. The reader may recall that in chapter 1 we made no direct assumption about the parent distribution(s) of the theoretical open variables. The justification for this leniency will now be apparent; all we require is that Fisher z formed from the transformed variables (Q, V, or Y) be sufficiently well behaved, and for the types of transformation involved this is usually a reasonable requirement. (When there is reason to suspect otherwise, no test should be attempted.)

Exercises

6.2 The sample correlations of quartz with microcline and of microcline with plagioclase for the data of exercises 5.4 and 6.1 are $r_{12} = -0.6466$ and $r_{23} = -0.6543$. Show (a) that these results are significant at the 0.01 level against the alternative that $\rho = 0$, and (b) that each fails of significance at the 0.50 level against the appropriate ρ_{ij} from exercise 6.1. Answers:

a. Reference to a table of the Student-t distribution shows that, for d.f. = 13, $t_{(0.01)} = 3.012$. From equation (6.2a), $t' = 3.056$ for r_{12} and 3.120 for r_{23}.

b. Reference to a table of the cumulative normal frequency distribution shows that $t_{(0.50)} = 0.674$. From equation (6.4), $t'' = 0.38$ for r_{12} and 0.33 for r_{23}.

Hence the sample correlations are highly significant against the conventional null hypothesis that $\rho = 0$ but fail entirely of significance against values of ρ derived from the null model of chapter 1. It is to be remembered, of course, that the critical values of t apply strictly only for bivariate normal distributions, and that, in addition, t'' is an approximation.

3. Closure Correlation in Reference Material G-2

Numerical calculations have so far been confined to exercises. Both because of the intrinsic interest of the data involved and because an extensive review of the work of the opening chapters seems appropriate, we now make an exception to this rule. The data shown in table 6.1 are modal analyses of 12 thin sections cut from a block of Bradford, Rhode Island, granite included in the current set of U.S. Geological Survey analytical reference materials under the designation G-2. To the extent that such a property can be tested by modal analysis, the Bradford block is satisfactorily uniform in composition. The most stringent test of (modal) sample variance is a comparison of entries in the last two lines of table 6.1. The observed variance for each mineral is not significantly different from the binomial counting error computed for the average count length of 2020 points; that is to say, differences between thin sections are too small for reliable detection with counts of this length. The largest difference is for plagioclase, and the quantity $F = (1.38/1.10)^2 = 1.574$ fails of significance at the 0.05 level, and almost at the 0.10 level, so there is little reason to suppose the sample variance contains anything but counting error.

There are nevertheless strong correlations between certain pairs of variables, specifically, $r_{13} = -0.7267$, $r_{23} = -0.7765$, $r_{24} = -0.5978$; the scatter diagrams of figure 1.1 show the data for the first two pairs. This is as clear an example as could be desired that strong correlation is not incompatible with uniformity in composition, for against the hypothesis that $\rho = 0$ the test described in the preceding section indicates that r_{13} and r_{23} are significant at the 0.01 and r_{24} at the 0.05 level.

To determine whether these correlations should be regarded as more than closure effects we first compute the elements of $\boldsymbol{\sigma}$ from equations (5.6) or (5.15); these are, in order, 0.5141, 1.7850, 4.2668, 0.4409, and 0.0927. We then calculate the closure correlations, using equation (6.1); the results of this calculation are shown in the column headed ρ in table 6.2.

Entering equation (6.4) with r and ρ from each line of the table, we obtain the entries in the column headed t''. The 0.10 point for t_α is 1.64, and since no t'' in the table exceeds this value, none of the sample correlations warrants rejection, at the 0.10 level, of the null hypothesis that the observed associations are closure effects.

Table 6.1 Modal Analyses of Twelve Thin Sections of Bradford, Rhode Island, Granite

Variable i	Quartz 1	Microcline 2	Plagioclase 3	Biotite 4	Others 5
	22.4	27.2	42.8	5.7	1.9
	22.8	26.9	42.0	6.4	1.9
	21.3	28.2	42.7	6.1	1.7
	21.2	26.4	44.0	6.5	1.9
	20.5	25.3	45.7	6.8	1.7
	20.8	28.2	43.5	5.1	2.4
	20.9	25.3	44.8	7.0	2.0
	22.0	27.3	41.5	7.0	2.2
	22.0	27.8	43.3	5.4	1.5
	20.3	25.5	45.6	6.1	2.5
	20.8	28.3	42.3	6.4	2.2
	21.7	28.0	42.5	5.4	2.4
$\bar{x}$	21.4	27.0	43.4	6.2	2.0
s	0.79	1.16	1.38	0.64	0.30
Binomial counting error	0.91	0.99	1.10	0.53	0.31

Source: U.S.G.S. reference sample G-2; data summarized in Chayes (1967).

Note: Both to conform with convention and to avoid a plethora of leading zeros, the entries in this table are percentages of the whole. Readers interested in checking the calculations should recall once more that the p of equations (4.10), (5.6), and (5.15) is a proportion, not a percentage—that is, $\bar{y} = 10^2p$. In calculations of this sort, incidentally, rounding error cumulates very rapidly, and it is well to carry as many decimals as possible. The entries in table 6.2 are rounded results based on calculations in which means and variances were carried, respectively, to the fourth and eighth digit past the decimal point.

Table 6.2 Sample Correlations (r) and Expected Closure Correlations (ρ) for the Data of Table 6.1

ij	r_{ij}	ρ_{ij}	t''_{ij}
12	+0.3548	−0.1212	1.48
13	−0.7267	−0.4395	1.35
14	−0.1445	−0.0641	0.24
15	−0.3383	+0.0019	1.06
23	−0.7765	−0.6865	0.59
24	−0.5978	−0.1491	1.62
25	+0.0622	−0.0637	0.38
34	+0.1534	−0.2982	1.39
35	−0.2572	−0.1536	0.32
45	−0.1621	−0.0338	0.39

4. Two Major Limitations of the Test

The conventional null hypothesis that $\rho = 0$ is clearly inappropriate for a test of departures from random association between any pair of major constituents in a closed array, but the alternative proposed here, although a considerable improvement, is far from ideal. We must now examine more closely the structure and meaning of the rather extensive calculations in which we have become involved.

As an entry to the problem, we have chosen parameters of **X** such that the means and variances of **Y** formed by the closure of **X** will be just those of the sample. In thus ignoring sampling variation of means and variances we are in effect assuming, for the purpose of our test, that these parameters of **Y** are known without error. This is manifestly unrealistic, but at the moment no practical alternative is available.

A second weakness of the test is that it is something of a bivariate sheep in the clothing of a multivariate wolf. In setting up the test the general apparatus appropriate to multivariate analysis is convenient, for it is clear that the means and variances of all variables in **X** influence both the variance of each variable and the covariance of each pair of variables in **Y**. Yet although the analysis uses some of the devices of multivariate algebra, the resulting test is not genuinely multivariate. Any individual correlation may be tested, and it might even be legitimate to test each one in turn, but no provision is made for a group or ensemble test. Such a test might be very useful, for it is certainly possible that although no individual correlation in a set warrants rejection of the null hypothesis, the correlations as a group are unlikely to have been generated entirely by closure. In the data of the preceding section, for instance, all ten correlations fail at the 0.10 level, but one is significant at the 0.15 level and three others are significant at the 0.20 level. This is hardly much cause for suspicion; if, however, all of the correlations had failed of significance at the 0.01 level but one of the ten had been significant at the 0.05 and three more at the 0.10 level, the case might have been quite otherwise.

5. The Interpretation of Negative Elements in σ

Except for occasional almost parenthetic specification that $\sigma_i^2 \geqslant 0$ for all i, we have so far not considered the possibility that for one or more of the variances the solution of equations (5.6) or (5.16) is negative, a perfectly permissible result. Variance is by definition positive, and if a solution of equations (5.6) or (5.16) does in fact yield some σ_i^2 which is negative, an **X** like that specified in the null model of chapter 1 is a numerical impossibility. We must then conclude that the null hypothesis based upon it is to be rejected. This is a group test of a kind, but it is an exceedingly crude one; instead of rejecting the hypothesis because the observed covariances are unlikely we are simply unable to make the test because the hypothetical open variances appear to be impossible.

From examination of equation (5.6) it is evident that, if no $p \geqslant \frac{1}{2}$, $\sigma_i^2 < 0$ if and only if

$$\frac{s_i^2}{p_i^2} < \frac{\sum_{j=1}^{m}\left(\frac{s_j^2}{1-2p_j}\right)}{\sum_{j=1}^{m}\left(\frac{p_j(1-p_j)}{1-2p_j}\right)}$$

In chapter 9 we shall see that, under circumstances fortunately rather common in practice, this rule continues to hold even when some $p > \frac{1}{2}$.

6. A Further Note on the Ternary Array

As shown in chapter 4, the correlations in any ternary array can be calculated exactly from the variances. Since in the test of closure correlation proposed here we purposely choose the means and variances of **X** so that var $(Y_i) = s_i^2$ for all i, it necessarily follows that if σ is free of negative elements we shall always have $\rho_{ij} = r_{ij}$ in any ternary array. There is of course no point in testing a null hypothesis that cannot possibly be rejected. The moral is clear; the test of closure correlation is inapplicable unless $m \geqslant 4$.

This injunction extends to the stratagem, described in chapter 4, of reducing an m-variable array to a ternary one in order to facilitate calculation of the expected correlations. The set of expectations obtained in this fashion will differ from the set found for the unreduced array. The reduced set will be exact and trivial; the unreduced set will be approximate but nontrivial.

7 The Effect of Ratio Formation on Regression

At each step of the preceding argument a slight shift of emphasis would have led to the formation of null values for the coefficients of regression rather than of correlation. In fact, the two developments could have been carried along simultaneously with very little additional work. Because of subtle but important interpretive differences, however, it is more convenient to discuss regression separately. In the text or exercises of the first three sections of this chapter null regression coefficients are derived for each pair of variables for which null correlations were derived in earlier chapters.

1. The Analytical Relation between Regression and Correlation Coefficients

The coefficient of regression of variable i on variable j, the slope of the least mean square line of best fit for i as a function of j, is the ratio of the covariance of i with j to the variance of j. Symbolically, for any sample,

$$b_{ij} = s_{ij}/s_j^2 \tag{7.1}$$

and, rotating subscripts,

$$b_{ji} = s_{ij}/s_i^2. \tag{7.2}$$

Similarly, the population regression coefficients are

$$\beta_{ij} = \sigma_{ij}/\sigma_j^2 \text{ and } \beta_{ji} = \sigma_{ij}/\sigma_i^2. \tag{7.3}$$

The sample statistics and parent parameters required for the calculation of observed and parent or null regression coefficients are thus exactly those needed for analogous correlation calculations. In fact, the signs of r, b, ρ, and β are those of the relevant covariance, and the absolute value of ρ (or r) is the geometric mean of $\beta_{ij}\beta_{ji}$

(or $b_{ij}b_{ji}$). It is thus a simple matter to calculate a pair of β's for each ρ found in the preceding chapters.

2. Null Regression Coefficients for the Simple Ratios

For the first three ratio pairings of table 2.1, for instance, we have that

a) if $Q_i = X_1/X_2$ and $Q_j = X_1$, then $\beta_{ij} \cong \mu_2{}^{-1}$ and

$$\beta_{ji} \cong \frac{\mu_2^3\sigma_1^2}{\mu_2^2\sigma_1^2 + \mu_1^2\sigma_2^2},$$

b) if $Q_i = X_1/X_2$ and $Q_j = X_2$, $\beta_{ij} \cong -\mu_1/\mu_2^2$ and

$$\beta_{ji} \cong \frac{-\mu_1\mu_2^2\sigma_2^2}{\mu_2^2\sigma_1^2 + \mu_1^2\sigma_2^2},$$

and

c) if $Q_1 = X_1/X_2$ and $Q_j = X_3/X_2$,

$$\beta_{ij} \cong \frac{\mu_1\mu_3\sigma_2^2}{\mu_2^2\sigma_3^2 + \mu_3^2\sigma_2^2} \text{ and}$$

$$\beta_{ji} \cong \frac{\mu_1\mu_3\sigma_2^2}{\mu_2^2\sigma_1^2 + \mu_1^2\sigma_2^2}.$$

In each case we merely substitute into $\sigma_{ij}/\sigma_i{}^2$ or $\sigma_{ij}/\sigma_j{}^2$ the values of σ_{ij} and $\sigma_i{}^2$ or $\sigma_j{}^2$ which we have already had to calculate in order to find ρ_{ij}. The procedure is straightforward, although, as will be noted in some of the following exercises, the results are sometimes rather impressively complex.

Exercises

Suggestion: Do at least one of the first three and at least one of the last three.

7.1. Using equations (2.18), (2.20), and (2.22), show that, if $Q_i = X_2/X_1$ and $Q_j = X_2/X_3$, $\beta_{ij} \cong \mu_3^2\sigma_2^2/\mu_1(\mu_3^2\sigma_2^2 + \mu_2^2\sigma_3^2)$; and $\beta_{ji} \cong \mu_1^3\sigma_2^2/\mu_3(\mu_1^2\sigma_2^2 + \mu_2^2\sigma_1^2)$.

7.2. Using equations (2.5), (2.20), and (2.25), show that, if $Q_i = X_1/X_2$ and $Q_j = X_2/X_3$, $\beta_{ij} = -\mu_1\mu_3^3\sigma_2^2/\mu_2^2(\mu_3^2\sigma_2^2 + \mu_2^2\sigma_3^2)$ and $\beta_{ji} \cong -\mu_1\mu_2^2\sigma_2^2/\mu_3(\mu_2^2\sigma_1^2 + \mu_1^2\sigma_2^2)$.

7.3. Using equations (3.2), (3.7), and (3.9), show that, if $Q_1 = X_1 + X_2$ and $Q_j = X_1$, $\beta_{ji} = \sigma_1^{-1}(\sigma_1^2 + \sigma_2^2)^{-1/2}$ and $\beta_{ij} = \sigma_1^2/(\sigma_1^2 + \sigma_2^2)^{3/2}$.

7.4. Given that $Q_i = X_1 + X_2$ and $Q_j = X_1 + X_3$, show that $\beta_{ji} = \sigma_1^2/(\sigma_1^2 + \sigma_2^2)$ and $\beta_{ij} = \sigma_1^2/(\sigma_1^2 + \sigma_3^2)$.

7.5. Given that $Q_i = X_1 + X_2$ and $Q_j = X_1/(X_1 + X_2)$, show that $\beta_{ij} \cong (\mu_1 + \mu_2)^2(\mu_2\sigma_1^2 - \mu_1\sigma_2^2)/(\mu_2^2\sigma_1^2 + \mu_1^2\sigma_2^2)$ and $\beta_{ji} \cong (\mu_2\sigma_1^2 - \mu_1\sigma_2^2)/(\mu_1 + \mu_2)^2(\sigma_1^2 + \sigma_2^2)$.

7.6. Given that $Q_i = Y_1 + Y_2$ and $Q_j = Y_1/(Y_1 + Y_2)$, show that $\beta_{ij} \cong (1 - \mu_1 - \mu_2)(\mu_2\sigma_1^2 - \mu_1\sigma_2^2)(\mu_1 + \mu_2)^2/(\mu_2^2\sigma_1^2 + \mu_1^2\sigma_2^2)$ and $\beta_{ji} \cong (1 - \mu_1 - \mu_2)(\mu_2\sigma_1^2 - \mu_1\sigma_2^2)/(\mu_1 + \mu_2)^2[(1 - 2\mu_1 - 2\mu_2)(\sigma_1^2 + \sigma_2^2) + (\mu_1 + \mu_2)^2\sigma_t{}^2]$.

3. Null Regression Coefficients between Ratios Whose Numerators are Common Elements in the Sum Which Is Their Common Denominator ($Y_i = X_i/T$, $Y_j = X_j/T$, $T = \Sigma X$)

This is the by now familiar case of the closed array, in which closure has been imposed on subsets drawn from an open parent characterized by zero covariances. The algebra is considerably simpler than in some of the exercises of the preceding section, but the situation is of such importance in petrography as to warrant separate discussion.

From the work of chapter 4 we have that

$$\text{var}(Y_i) \cong p_i{}^2\sigma_t{}^2 + (1 - 2p_i)\sigma_i{}^2, \qquad (4.6)$$

$$\text{var}(Y_j) \cong p_j{}^2\sigma_t{}^2 + (1 - 2p_j)\sigma_j{}^2, \qquad (4.8)$$

and

$$\text{cov}(Y_i, Y_j) \cong p_i p_j \sigma_t{}^2 - p_i\sigma_j{}^2 - p_j\sigma_i{}^2, \qquad (4.9)$$

so that, obviously,

$$\beta_{ij} \cong (p_i p_j \sigma_t{}^2 - p_i\sigma_j{}^2 - p_j\sigma_i{}^2)/[p_j{}^2\sigma_t{}^2 + (1 - 2p_j)\sigma_j{}^2], \qquad (7.4)$$

$$\beta_{ji} \cong (p_i p_j \sigma_t{}^2 - p_i\sigma_j{}^2 - p_j\sigma_i{}^2)/(p_i{}^2\sigma_t{}^2 + (1 - 2p_i)\sigma_i{}^2), \qquad (7.5)$$

and

$$\rho_{ij}^2 = \beta_{ij}\beta_{ji} \cong \frac{(p_i p_j \sigma_t^2 - p_i \sigma_j^2 - p_j \sigma_i^2)^2}{(p_i^2 \sigma_t^2 + (1 - 2p_i)\sigma_i^2)(p_j^2 \sigma_t^2 + (1 - 2p_j)\sigma_j^2)} \quad (7.5a)$$

If we work without assumption about the distribution of **X**, we estimate $\boldsymbol{\beta}$ only after first approximating **p** and $\boldsymbol{\sigma}$ in the fashion described in chapter 5; that is to say, we assign means and variances to the elements of **X** such that, on closure, the elements of **Y** will have the same means and variances as the analogous elements of the sample vector. The regression coefficients between any pair of variables in **X**, say X_i and X_j, are of course zero, since by our initial assumption $\sigma_{ij} = 0$. But the regression coefficients between the analogous variables in **Y**, and hence to be anticipated, on our null hypothesis, between the ith and jth sample variables, are in general not zero; rather, to first-order approximation they are the results shown in equations (7.4) and (7.5).

Recalling the simplifying assumptions about distribution already employed in the discussion of closure correlation, let us suppose first that means and variances are homogeneous in **X**, that is, that $p_i = 1/m$ and $\sigma_i^2 = \theta$ for all i. Since $\sigma_t^2 = \Sigma\sigma_i^2$, we then also have that $\sigma_t^2 = m\theta$. Substituting these values for $\sigma_i^2, \sigma_j^2, p_i, p_j$, and σ_t^2 in equations (7.4) and (7.5), we readily obtain $\beta_{ij} = \beta_{ji} \cong -1/(m-1)$, the result previously found for ρ_{ij} under the same assumptions. This is as it should be, for if $\sigma_i^2 = \sigma_j^2$, the quantities $\rho_{ij} = \sigma_{ij}/\sigma_i\sigma_j$ and $\beta_{ij} = \beta_{ji} = \sigma_{ij}/\sigma_j^2 = \sigma_{ij}/\sigma_i^2$ are obviously identical, and, by definition, the absolute value of ρ_{ij} is the geometric mean of β_{ij} and β_{ji}.

Supposing next that variances are proportional to means in **X**, we have that $\sigma_i^2 = kp_i$ for all i, from which $\sigma_t^2 = \Sigma\sigma^2 = k$. By substitution of these values for σ_i^2, σ_j^2, and σ_t^2, equations (7.4) and (7.5) are easily reduced to $\beta_{ij} \cong -p_j/(1-p_i)$ and $\beta_{ji} \cong -p_i/(1-p_j)$. The closure correlation if $\sigma_i^2 = kp_i$ for all i, given in equation (4.14), is $\rho_{ij} \cong -\{p_i p_j/(1-p_i)(1-p_j)\}^{1/2}$, so that, as it should, $\rho_{ij}^2 = \beta_{ij}\beta_{ji}$.

In computing β_{ij}, β_{ji} for an actual sample, it is of course not necessary to calculate var (Y_i) or var (Y_j) at all, since by definition they are equal to the sample variances. In analogy with equation (6.1) we thus have

$$\beta_{ij} = \text{cov}\,(Y_i, Y_j)/\text{var}\,(Y_j) \cong (\bar{y}_i\bar{y}_j\sigma_t^2 - \bar{y}_i\sigma_j^2 - \bar{y}_j\sigma_i^2)/s_j^2$$

and, similarly, (7.6)

$$\beta_{ji} = \text{cov}\,(Y_i, Y_j)/\text{var}\,(Y_i) \cong (\bar{y}_i\bar{y}_j\sigma_t^2 - \bar{y}_i\sigma_j^2 - \bar{y}_j\sigma_i^2)/s_i^2.$$

Exercises

7.7. From the data display in exercise 5.4, compute, by substitution in equation (7.6), $\beta_{12} \cong -0.4267$, $\beta_{21} \cong -0.7831$, $\beta_{23} \cong -0.7871$, and $\beta_{32} \cong -0.4524$.

7.8. From the information given in table 6.1 and the accompanying discussion of reference material G-2, compute the null regression coefficients for quartz and plagioclase ($\beta_{13} \cong -0.2504$, $\beta_{31} \cong -0.7640$), for microcline and plagioclase ($\beta_{23} \cong -0.5748$, $\beta_{32} \cong -0.8136$), and for microcline and biotite ($\beta_{24} \cong -0.2707$, $\beta_{42} \cong -0.0824$).

If you now check by comparing each $(\beta_{ij}\beta_{ji})^{1/2}$ with $|\rho_{ij}|$ from table 6.2, the agreement will extend only to three places for two of the values. The entries in table 6.2 were calculated with four significant digits in means and eight in variances while in this problem you are instructed to use data from table 6.1, in which means are listed to only three significant figures and standard deviations to only two or three.

4. Testing the Regression Coefficient[1]

The sample standard deviation of the regression coefficient of X_i on X_j, that is, of the dependent variable X_i, as a function of the independent variable X_j, is

$$s_{b_{ij}} = \frac{s_i}{s_j}\sqrt{\frac{1 - r_{ij}^2}{n - 2}}. \tag{7.7}$$

If the sample is drawn from a bivariate normal parent the quantity $(b_{ij} - \beta_{ij})/s_{b_{ij}}$—where b and β are the sample and parent coefficients of regression—is distributed as Student's t with $(n - 2)$ degrees of freedom and the quantity $(b_{ij} - \beta_{ij})^2/s_{b_{ij}}^2$ is distributed as Snedecor's F with one degree of freedom in the numerator and $(n - 2)$ degrees of freedom in the denominator. (Analogous results may be obtained for b_{ji} by rotating subscripts.)

If $\beta = 0$, as in the conventional null hypothesis, a test of the significance of an observed regression coefficient is in fact a test of the significance of the sample r against the hypothesis that $\rho = 0$, for

$$\frac{b_{ij}}{s_{b_{ij}}} = \frac{\frac{s_i}{s_j} r_{ij}}{\frac{s_i}{s_j}\sqrt{\frac{1 - r^2}{n - 2}}} = r_{ij}\sqrt{(n - 2)/(1 - r_{ij}^2)}, \tag{7.8}$$

[1] Readers familiar with elementary regression analysis will find this section a largely unnecessary review.

exactly as in equation (6.2a). Against $\rho \neq 0$ the test is inapplicable, and, as we have seen, it is convenient to fall back on the Fisher z transformation. No such complication arises in connection with the testing of b against the alternative that $\beta \neq 0$.

Despite its advantage in this respect, at the present point in our discussion we cannot make much use of the regression coefficient, for in order to do so we must be able to specify one variable as dependent and the other as independent. To petrologists accustomed to thinking in terms of deterministic models this objection often seems niggling, for if in fact $X_i = f(X_j)$, then of course $X_j = g(X_i)$, and we may state either variable as a function of the other; the traces of the two functions will be identical, and the relation between them purely algebraic. This is unfortunately not true of regression functions. Obviously we may "solve"

$$\hat{X}_i = a_j + b_{ij}X_j \tag{7.9}$$

for X_j and find out what values of X_j would give particular estimates of X_i, but in general if we substitute $\hat{X}_i$ from equation (7.9) for X_i in

$$\hat{X}_j = a_i + b_{ji}X_i, \tag{7.10}$$

we shall find that $\hat{X}_j \neq X_j$, and, except for a small region in the immediate vicinity of the point $(\bar{x}_i, \bar{x}_j)$, the difference will be larger than the error of estimate unless $|r_{ij}| \cong 1$. We can always compute equations (7.9) and (7.10), but we cannot properly decide which one to use unless we are able to specify, a priori, one of the variables as independent, and the other as dependent upon it. In our random vector, **X**, the variables are by definition uncorrelated, and, from the underlying assumptions, no one is dependent on any other. The operations of ratio formation generate interdependence between some of the variables, and closure produces some interdependence between all of them, but in neither case can we say that after the transformation one of the variables is dependent on another which is independent of it. We shall thus usually be able to carry the regression analysis only to the point of deciding whether or not the b's differ significantly from the analogous β's. As we have just seen, in equation (7.8) and the accompanying discussion, however, this is nothing more than we have already discovered by testing r_{ij} against ρ_{ij}. In the interpretation of variation diagrams, petrographers often proceed graphically as if they were engaged in a regression analysis, but in most instances the only valid inferences to be drawn concern correlation, not regression.

Exercises

7.9. From the definitions of β, b, and s_b, show that $t_{ij} = t_{ji} = (r_{ij} - \rho_{ij})\sqrt{[(n-2)/(1-r_{ij}^2)]}$. Suggestion: Substitute $b_{ij} - \beta_{ij} = (s_i/s_j)(r_{ij} - \rho_{ij})$ and $s_{b_{ij}} = (s_i/s_j)\sqrt{[(1-r_{ij}^2)/(n-2)]}$ into $t_{ij} = (b_{ij} - \beta_{ij})/s_{b_{ij}}$. Note that t will be the same whether found for b_{ij} or b_{ji}, so that it is necessary to test only one of any pair of regression coefficients.

7.10. From data given in the discussion of U.S. Geological Survey reference material G-2 (chap. 6), show that b_{13} fails of significance at the 0.2 level. Answer: $b_{13} = (0.79)(-0.7267)/1.38 = -0.4160$, $\beta_{13} = (0.79)(-0.4395)/1.38 = -0.2516$, $s_{b_{13}} = (0.79/1.38)(0.4719/10)^{1/2} = 0.1244$, so that $t_{13} = (b_{13} - \beta_{13})/s_{b_{13}} = -1.322$, whereas $Pr(|t| < 1.37 \mid n = 12) > 0.8$.
Of course, the same result can be reached, with much less arithmetic, from the form of t found in exercise 7.9, that is, $t_{13} = (-0.7267 + 0.4395)\sqrt{(10/0.4719)} = -1.322$. Thus the test of the significance of regression is a function only of correlation. In fact, it may be shown that the proportion of the total sum of squares "accounted for" by regression is simply r^2.

5. Alternative Forms of Regression Analysis

In many fields, perhaps most particularly in biology, interest often centers on the joint variation of variables whose relation to each other is one of interdependence rather than the strict dependence-independence required in the usual formulation. In those natural sciences strongly influenced by physics and chemistry, on the other hand, it is common to suppose that there is an underlying functional relation between the variables whose association is of interest, and that the values recorded in any particular set of data exhibit this functional relation distorted or made inexact by errors of measurement.

There are profound differences between the concepts of statistical interdependence and basically nonstatistical functional relation, but a situation or model governed by either usually brings the same properties of conventional regression analysis into question. The identification of one variable as independent of the other is manifestly impossible if the relation is one of interdependence, and frequently so if it is presumed to be functional.[2] In both situations, the announced values for each variable are subject to error. Perhaps most important, in both situations the existence of two quite different regression equations that fit the data equally well seems at best unrealistic and at worst absurd.

In recent years there has been considerable progress toward a more realistic statistical analysis of data characterized by an underlying functional relation between the variables of interest (for a summary, see Kendall and Stuart, chap. 29). Much contemporary speculation about petrographic variation diagrams does indeed presume a functional relation between the variables, and it is hardly to be doubted that statistical developments in this field will one day find broad application in petrology. In the present discussion, however, they are largely irrelevant, for we are concerned only with relations that can be imposed upon initially uncorrelated random variables by the operations of closure and ratio formation. Functional relation cannot be generated in this fashion.

The relation which can be imposed by these operations is one of interdependence and it has been suggested (see, for instance, Kermack and Haldane, or Kruskal) that for interdependent variables the appropriate regression function is the so-called reduced major axis or line of organic correlation, which is very easily found.

For any pair of variables in **Y**, for instance, it is simply

$$\hat{Y}_i = \gamma_j + \beta_j Y_j, \tag{7.11}$$

or

$$\hat{Y}_j = \gamma_i + \beta_i Y_i, \tag{7.12}$$

where

$$\gamma_j \equiv -\beta_j \gamma_i \cong \bar{y}_i - \beta_j \bar{y}_j,$$

and (7.13)

$$|\beta_j| \equiv |\beta_i^{-1}| \cong s_i/s_j.$$

The sign of each β is that of cov (Y_i, Y_j), the concept of interrelatedness being meaningless except when cov $(Y_i, Y_j) \neq 0$. In view of

---

[2] This is certainly not true of all functionally related variables. No one will have difficulty deciding which of the pair (gradient, rate of flow) is the independent variable, for instance, but which of the variables in any of the standard petrographic variation diagrams is properly regarded as independent of the others?

definitions in equation (7.13) and this sign convention, the loci of points satisfying equations (7.11) and (7.12) are identical. Thus, there is only one line of organic correlation; it passes through the point $(\bar{y}_i, \bar{y}_j)$, like the conventional regression lines discussed in the preceding section, and lies in the acute angle between them.

From the way in which **Y** is defined—that is, so that to the limits of the approximation its means and variances are identical with those of the sample—no test of the significance of an observed b is possible, for $|\beta_j| \cong s_i/s_j = |b_j|$, and similarly for variable i. But no test is necessary; in practice equation (7.11) is to be computed or drawn only if a prior test of r_{ij}, or of the array as a whole, indicates that the association cannot be accounted for by closure. (For pedagogic purposes, an exception to this rule is made in exercise 7.13 below.)

Exercises

7.11 The null regression equations for any pair of variables in **Y**, say Y_i, Y_j, are, following our established notation, $\hat{Y}_j = \alpha_i + B_{ji}Y_i$, and $\hat{Y}_i = \alpha_j + \beta_{ij}Y_j$, where $\alpha_i = \bar{y}_j - \beta_{ji}\bar{y}_i$. These lines intersect at the point $(\bar{y}_i, \bar{y}_j)$, since if $Y_i = \bar{y}_i$, $\hat{Y}_j = \bar{y}_j$, and if $Y_j = \bar{y}_j$, $\hat{Y}_i = \bar{y}_i$. Denoting the slope angles by $\omega_i = \tan^{-1}\beta_{ji}$, $\omega_j = \tan^{-1}\beta_{ij}$, show that the "null" angle of intersection is

$$\hat{\theta} = \tan^{-1}\left(\frac{1 - \tan\omega_i \tan\omega_j}{\tan\omega_i + \tan\omega_j}\right) = \frac{s_i s_j}{s_i^2 + s_j^2} \times \frac{1 - \rho_{ij}^2}{\rho_{ij}} .$$

(To find the observed angle of intersection, θ, replace ρ_{ij} by r_{ij} in this expression.)

7.12 Show that the line of organic correlation also passes through the point $(\bar{y}_i, \bar{y}_j)$, and that it lies in the angles $\hat{\theta}, \theta$, defined in exercise 7.11.

Suggestions: To prove that $(\bar{y}_i, \bar{y}_j)$ is a point on the reduced major axis, show that it is a solution of equation (7.11). To show that the reduced major axis lies within $\hat{\theta}$, note first that $|\beta_{ij}| = |\rho_{ij}\beta_j|$ and $|\beta_{ji}| = |\rho_{ij}\beta_i|$. Next, recall that $|\rho_{ij}| < 1$ and that ρ_{ij} and all the β's take their sign from the covariance. Hence the slopes must always be such that $|\beta_j| > |\beta_{ij}|$ and $|\beta_i| > |\beta_{ji}|$, each angle being read against the appropriate reference axis. This completes the proof as regards $\hat{\theta}$; the relation also holds for θ, as may be seen by replacing each ρ_{ij} with r_{ij}.

7.13 From the data of table 6.1 and relations of equations (7.11) and (7.13), compute the lines of organic correlation $\hat{Y}_1 = 46.2 - 0.582Y_3$ and $\hat{Y}_2 = 63.5 - 0.841Y_3$. Plot these on scatter diagrams prepared from table 6.1, or add them directly to fig. 1.1. They are excellent examples of rather convincing "trend lines" that could have been generated entirely by closure, for we have already shown that r_{13} and r_{23} fail of significance against ρ_{13} and ρ_{23} (see table 6.2 and accompanying discussion).

7.14 If $\kappa_i = \tan^{-1} \beta_i, \kappa_j = \tan^{-1} \beta_j, \omega_i = \tan^{-1} \beta_{ji}$, and $\omega_j = \tan^{-1} \beta_{ij}$, show that the angles between the line of organic correlation and the regression lines are $\hat{\phi}_i = \kappa_i - \omega_i = \tan^{-1} (\beta_i - \beta_{ji})/(1 + \beta_j\beta_{ji}), \hat{\phi}_j = \kappa_j - \omega_j = \tan^{-1} (\beta_j - \beta_{ij})/(1 + \beta_i\beta_{ij})$.

Using the definitions of the β's, reduce these further, to obtain

$$\hat{\phi}_i = \tan^{-1} \left[\pm \frac{s_i s_j(1 \mp \rho_{ij})}{s_i{}^2 \pm s_j{}^2\rho_{ij}} \right],$$

and

$$\hat{\phi}_j = \tan^{-1} \left[\pm \frac{s_i s_j(1 \mp \rho_{ij})}{s_j{}^2 \pm s_i{}^2\rho_{ij}} \right],$$

the upper signs applying when $\rho_{ij} > 0$, the lower when $\rho_{ij} < 0$. Note than in general $\hat{\phi}_i \neq \hat{\phi}_j$; the reduced major axis bisects the angle between the regression lines if and only if var $(Y_i) =$ var (Y_j).

Readers who enjoy trigonometrical algebra may wish to check these results by showing that, in fact,

$$\tan (\hat{\phi}_i + \hat{\phi}_j) = \frac{s_i s_j}{s_i{}^2 + s_j{}^2} \times \frac{1 - \rho_{ij}{}^2}{\rho_{ij}} = \tan \hat{\theta},$$

where $\hat{\theta}$ is, as defined in exercise 7.11, the angle between the regression lines.

7.15 Applying the results of exercises 7.11 and 7.13 to the data of tables 6.1 and 6.2, show that for variables 2 and 3 (microcline and plagioclase), the angle between the regression lines is $-14.2°$ as compared to a null value of $-20.8°$, and that $\hat{\phi}_3$ is -10.1 whereas ϕ_3 calculated directly from the sample statistics is $-6.95°$. (To compute ϕ_3 instead of $\hat{\phi}_3$, enter either formula of exercise 7.13 with r_{ij} instead of ρ_{ij}. Although, in our null model, there is no difference between b_i and β_i, there is a difference between ϕ_3 and $\hat{\phi}_3$; this merely reflects the difference between b_{ji} and β_{ji}.)

7.16 Show that, for constant means and variances, the observed reduced major axis is the line toward which the regression lines rotate as correlation increases in strength.

Suggestion: It has already been shown in example 7.12 that the three lines intersect at the point $(\bar{y}_i, \bar{y}_j)$, so only the slopes must be considered. The slopes of the regression lines may be written as $b_{ji} = r_{ij}s_j/s_i$ and $b_{ij} = r_{ij}s_i/s_j$. Hence, as $|r_{ij}| \to 1$, $b_{ji} \to \text{sgn}(s_{ij})\, s_j/s_i$, and $b_{ij} \to \text{sgn}(s_{ij})\, s_i/s_j$. But these limiting values for b_{ji} and b_{ij} are precisely the definitions of b_i and b_j.

8 The Remaining-Space Transformation and the Niggli Variation Diagram

Graphical data reductions and interpretations based upon them have long been of central importance in conventional chemical petrography. Such schemes permit no direct study of the multivariate systems which are our chief concern, but the temptation to rely upon them when we are dealing with few enough variables is very strong. Even here, however, as the work of the last four chapters shows, conclusions about the significance of associations between variables are suspect because in the inspection of a scatter diagram no systematic account can be taken of closure effects.

There are two possible escapes from this dilemma; either we must abandon graphical analysis or we must develop projections which have the effect of reducing expected closure correlation. In the first part of this chapter we propose a projection of closed data subjected to the conceptually simple but so far unused "remaining-space" transformation, a transformation that should often materially reduce the expected closure effect. This leads directly to a discussion of the similar but widely used Niggli-number transformation, and the chapter concludes with a demonstration that, to the level of approximation used here, null correlations between Niggli numbers are identical with those between analogous remaining-space variables. The whole argument—unfortunately, as it turns out—deals only with closed data arrays for which parent populations of open uncorrelated variables are in fact numerically possible.

1. Rationale of the Remaining-Space Transformation

We may suppose that the immediate parent from which the items in a sample are drawn occupies some finite volume, and that in this volume the variables must compete with each other for space. The volume in question may be a hand specimen, an outcrop, a lava flow, or a batholith. More generally, the immediate parent may also be conceived as **Y**, the parent closed array of preceding chapters, formed by closure of a theoretical open array, **X**, characterized by zero covariance.

Initially we ask merely whether there is some nonrandom association between the proportions of variables 1 and j, $2 \leq j \leq m$, in samples drawn from the parent volume (or from **Y**). In a scatter diagram of the sample values for any pair of variables, however, the effects of closure and the nonrandom effects, if any, generated by interdependence of the sort presumed in most petrographic hypotheses are inextricably confounded. Our apparently simple question is not at all simple; if we persist in relying entirely on graphical analysis it is in fact unanswerable.

Rephrasing the question, we may ask whether there is some nonrandom association between the proportion of variable 1 observed in samples of the whole space and the proportion of variable j in the space not occupied by variable 1 in these samples, that is, we replace Y_1 and Y_j by $V_1 = Y_1, V_j = Y_j/(1 - Y_1), 1 < j \leq m$. (In the present context the change of name of Y_1 is an unnecessary formality, but the change of all variable names when any one is transformed greatly simplifies subscript notation in the considerably more complicated transformations basic to the argument of the next chapter.)

At first glance it would seem that this simple transformation completely eliminates the closure effect, for in the absence of substantive relationship between them why should there be anything but zero covariance between one proportion and the ratio of a second to the sum of the remainder? But this escape from the problem is more apparent than real. What sound reason can we advance for ignoring the curious fact that in each item of the sample the variables from which the V's are estimated always sum to unity?

The expected correlation between V_1 and V_j if the V's are estimated from samples drawn from our parent or theoretical closed array, Y, is in general not zero, though it will often be closer to zero than that between Y_1 and Y_j. We next approximate this new null correlation in the by now familiar way, using the notation established in previous chapters.

2. The Remaining-Space Transformation

We require approximations of $E(V_1)$, $E(V_j)$, var (V_1), var (V_j), and cov (V_1, V_j). The mean, deviation, and variance of V_1 are, of course, exactly those of Y_1, that is,

$$E(V_1) \cong \mu_1/\tau = p_1, \tag{8.1}$$

$$\Delta_1 \cong (\delta_1 - p_1\delta_t)/\tau, \tag{8.2}$$

$$\text{var}\ (V_1) \cong (p_1^2\sigma_t{}^2 + (1 - 2p_1)\sigma_1^2)/\tau^2. \tag{8.3}$$

For the jth element of **V** we have

$$V_j = \frac{Y_j}{1 - Y_1} = \frac{X_j/T}{1 - (X_1/T)} = \frac{X_j}{T - X_1} = \frac{\mu_j + \delta_j}{(\tau - \mu_1) + (\delta_t - \delta_1)}$$

$$\cong \frac{\mu_j + \delta_j}{\tau - \mu_1}\left(1 + \frac{\delta_1 - \delta_t}{\tau - \mu_1}\right)$$

$$\cong \frac{p_j}{1 - p_1} + \frac{\delta_j}{\tau - \mu_1} + \frac{p_j}{1 - p_1}\left(\frac{\delta_1 - \delta_t}{\tau - \mu_1}\right). \qquad (8.4)$$

Thus,

$$E(V_j) \cong \frac{p_j}{1 - p_1}, \qquad (8.5)$$

and

$$\Delta_j = V_j - E(V_j) \cong \frac{1}{\tau - \mu_1}\left[\delta_j + \frac{p_j}{1 - p_1}(\delta_1 - \delta_t)\right]. \qquad (8.6)$$

The variance of V_j is the expectation of the square of equation (8.6), and the covariance of V_j with V_1 is the expectation of the product of equations (8.6) and (8.2). To reach the variance we thus require the expectation of

$$\Delta_j{}^2 \cong \frac{1}{(\tau - \mu_1)^2}\left[\delta_j{}^2 + 2\frac{p_j}{1 - p_1}(\delta_1\delta_j - \delta_j\delta_t) + \left(\frac{p_j}{1 - p_1}\right)^2(\delta_1^2 - 2\delta_1\delta_t + \delta_t{}^2)\right],$$

or

$$\text{var}(V_j) \cong \frac{1}{\omega^2}[q_j{}^2(\sigma_t{}^2 - \sigma_1^2) + (1 - 2q_j)\sigma_j{}^2], \qquad (8.7)$$

where, for convenience, we denote $\tau - \mu_1$ by ω, and $p_j/(1 - p_1)$ by q_j. The covariance is the expectation of

$$\Delta_1\Delta_j \cong \frac{1}{\tau\omega}(\delta_1 - p_1\delta_t)[\delta_j + q_j(\delta_1 - \delta_t)]$$

$$\cong \frac{1}{\tau\omega}[\delta_1\delta_j - p_1\delta_j\delta_t + q_j(\delta_1^2 - \delta_1\delta_t - p_1\delta_1\delta_t + p_1\delta_t{}^2)],$$

or

$$\text{cov}\,(V_1, V_j) \cong \frac{1}{\tau\omega}\,[p_1 q_j \sigma_\omega^2 - p_1 \sigma_j{}^2], \tag{8.8}$$

where $\sigma_\omega^2 = \sigma_t{}^2 - \sigma_1^2$.

Combining equations (8.8), (8.7), and (8.3), the null correlation between V_1 and V_j is

$$\rho_{V_1 V_j} \cong \frac{p_1 q_j \sigma_\omega^2 - p_1 \sigma_j{}^2}{\sqrt{[p_1^2 \sigma_t{}^2 + (1 - 2p_1)\sigma_1^2][q_j{}^2 \sigma_\omega^2 + (1 - 2q_j)\sigma_j{}^2]}}, \tag{8.9}$$

as compared to

$$\rho_{1j} \cong \frac{p_1 p_j \sigma_t{}^2 - p_j \sigma_1^2 - p_1 \sigma_j{}^2}{\sqrt{[p_1^2 \sigma_t{}^2 + (1 - 2p_1)\sigma_1^2][p_j{}^2 \sigma_t{}^2 + (1 - 2p_j)\sigma_j{}^2]}} \tag{4.10}$$

for that between Y_1 and Y_j.

In practical work we shall usually be tempted to resort to a transformation of this sort because of strong negative correlation between a major variable (1) and the others, much of it possibly attributable to closure. It is therefore of interest that $\text{cov}\,(V_1, V_j) > \text{cov}\,(Y_1, Y_j)$. The proof is surprisingly simple. The difference between the covariances is

$$\frac{1}{\tau\omega}\,[p_1 q_j \sigma_\omega^2 - p_1 \sigma_j{}^2] - \frac{1}{\tau^2}\,[p_1 p_j \sigma_t{}^2 - p_j \sigma_1^2 - p_1 \sigma_j{}^2], \tag{8.10}$$

and, since $1/\tau\omega > 1/\tau^2$, it will be sufficient to show that the term in the first bracket is larger than that in the second. The difference between the bracketed terms is

$$\begin{aligned} D &= p_1 q_j \sigma_t{}^2 - p_1 q_j \sigma_1^2 - p_1 \sigma_j{}^2 - p_1 p_j \sigma_t{}^2 + p_j \sigma_1^2 + p_1 \sigma_j{}^2 \\ &= p_1 (q_j - p_j) \sigma_\omega^2 + p_j (1 - p_1) \sigma_1^2 \\ &= q_j [p_1^2 \sigma_\omega^2 + (1 - p_1)^2 \sigma_1^2] \\ &= q_j \ \text{var}\,(Y_1) > 0, \end{aligned} \tag{8.11}$$

which completes the proof. Thus, the transformation will always increment the covariance of variables 1 and j, and since this covariance is usually a rather large negative number the usual effect of the transformation will be to shift it toward zero.

The effect of the transformation on the variance of variable j is not nearly so straightforward. Nevertheless, for the practical applications envisaged, in which $p_1 > 0.5$ and $p_j \ll 0.5$, it may be shown that ordinarily var $(V_j) >$ var (Y_j). From earlier work, we have for these variances

$$\text{var}\ (V_j) = \frac{1}{\omega^2}\ [q_j{}^2\sigma_\omega^2 + (1 - 2q_j)\sigma_j{}^2]$$

and (8.12)

$$\text{var}\ (Y_j) = \frac{1}{\tau^2}\ [p_j{}^2\sigma_t{}^2 + (1 - 2p_j)\sigma_j{}^2].$$

After elimination of the ω's, τ's, and q's, the ratio of the two variances is

$$\frac{\text{var}\ (V_j)}{\text{var}\ (Y_j)} = \frac{1}{(1 - p_1)^4}\left[\frac{p_j{}^2\sigma_\omega^2 + (1 - p_1)(1 - p_1 - 2p_j)\sigma_j{}^2}{p_j{}^2\sigma_t{}^2 + (1 - 2p_j)\sigma_j{}^2}\right] \tag{8.13}$$

The terms in $\sigma_j{}^2$ will ordinarily be small in relation to those in σ_ω^2 and $\sigma_t{}^2$, so whether or not var $(V_j) >$ var (Y_j) will nearly always depend on whether or not

$$\left(\frac{1}{(1 - p_1)^4}\right)\left(\frac{\sigma_\omega^2}{\sigma_\omega^2 + \sigma_1^2}\right) > 1,$$

or, equivalently,

$$\frac{\sigma_1^2}{\sigma_\omega^2} < \frac{1 - (1 - p_1)^4}{(1 - p_1)^4}\ . \tag{8.14}$$

For $p_1 > 1/2$, a condition almost invariably satisfied in Harker arrays, we may anticipate that var $(V_j) >$ var (Y_j) providing only that the open variance of the eliminated variable is not more than 15 times larger than the sum of the other open variances. This hardly seems a stringent requirement. Unless a σ free of negative elements can be calculated from the data, however, the actual ratio of σ_1^2 to σ_ω^2 cannot be found. It is therefore of some interest that in the numerous transformations so far calculated no exception to the rule that var $(V_j) >$ var (Y_j) has been encountered.

The usual effect of the transformation is thus to reduce the strength of the negative correlation between variables 1 and j; this comes about because the expected (negative) covariance always receives a positive increment, the expected variance of variable 1 is unchanged, and the expected variance of variable j is usually enlarged. Ordinarily, then, graphical appraisal of the strength of association of variables V_1 and V_j will be more reliable than that of Y_1 and Y_j, for the reason that the null correlation between V_1 and V_j in the absence of substantive effects is weaker than that between Y_1 and Y_j. Indeed, it may often be small enough so that in small samples the graphical appraisal is not substantially different from that in the standard situation, in which no closure correlation is involved.

Exercises

8. 1 The elements of σ for the data of table 6. 1 are listed in section 3 of chapter 6. Using these and the data of the table, find the null correlation between V_1 and V_2. Compute the sample correlation between these variables. Is the difference between sample and null correlation significant at the 0. 01 level?

8. 2 Show that if variances are proportional to means in **X**, var $(V_j)/$ var $(Y_j) > 1$ if and only if

$$p_j < \left\{\frac{1-(1-p_1)^2}{1-(1-p_1)^3}\right\}(1-p_1),$$

and construct a table or graph of p_j (max) as a function of p_1 for $0 < p_1 < 1$.

Hint: If variances are proportional to means in **X**, then $\sigma_t^2 \sim \tau$ and $\sigma_\omega^2 \sim \omega$. With these substitutions, the ratio var $(V_j)/$var (Y_j) readily reduces to

$$\frac{1}{(1-p_1)^3}\left[1-\frac{p_1}{1-p_j}\right],$$

which is greater than unity only when p_j satisfies the stated condition.

8. 3 If $Y_1 = SiO_2$, note that in the region of interest in most oxide variation diagrams, that is, $0.5 < p_1 < 0.7$ and $p_j < 0.25$ for all $j \neq 1$, the remaining-space transformation will enlarge the expected variance of Y_j if variances are proportional to means in **X**. With $p_1 = 0.7$ in the inequality stated in exercise 8. 2, p_j (max) is 0. 28, whereas with $p_1 = 0.5$, p_j (max) is 0. 43.

3. The Niggli-Number Transformation

We have previously supposed that the elements of **Y** are either volume or weight proportions, but for the discussion of Niggli numbers it is convenient to consider them molar proportions. Denoting SiO_2 by subscript 1, the Niggli numbers are then sums of quantities of the form $N_j = Y_j/\sum_{k=2}^{m}(Y_k)$. For $j > 1$ any N_j is thus a remaining-space variable whose mean, deviation, and variance are precisely those of V_j in the preceding section, that is,

$$E(N_j) \cong q_j,$$

$$\Delta_j \cong \frac{1}{\omega}[\delta_j + q_j(\delta_1 - \delta_t)], \qquad (8.15)$$

and

$$\text{var}\,(N_j) \cong \frac{1}{\omega^2}[q_j{}^2\sigma_\omega^2 + (1 - 2q_j)\sigma_j{}^2].$$

N_1, however, is not simply Y_1 under a new name, as was the case with V_1, for in the Niggli transformation the molar proportion of SiO_2 is also divided by the sum of the other molar proportions. To obtain the necessary statistics for N_1 we therefore merely substitute 1 for j in the first two lines of equation (8. 15).

The second line becomes

$$\Delta_1 = \frac{1}{\omega}[\delta_1 + q_1(\delta_1 - \delta_t)], \qquad (8.16)$$

and its square is

$$\Delta_1^2 \cong \frac{1}{\omega^2}[\delta_1^2 + 2q_1(\delta_1^2 - \delta_1\delta_t) + q_1^2(\delta_1^2 - 2\delta_1\delta_t + \delta_t{}^2)].$$

The variance of N_1 is the expectation of this square; the expectation of $(\delta_1^2 - \delta_1\delta_t)$ being zero, we are left with

$$\text{var}\,(N_1) \cong \frac{1}{\omega^2}[\sigma_1^2 + q_1^2(\sigma_t{}^2 - \sigma_1^2)]$$

$$\cong \frac{1}{\omega^2}[q_1^2\sigma_t{}^2 + (1 - q_1^2)\sigma_1^2] \qquad (8.17)$$

$$\cong \frac{1}{(1 - p_1)^4\tau^2}[p_1^2\sigma_t{}^2 + (1 - 2p_1)\sigma_1^2].$$

The covariance of N_1 with N_j is the expectation of

$$\Delta_1\Delta_j \cong \frac{1}{\omega^2}[\delta_1 + q_1(\delta_1 - \delta_t)][\delta_j + q_j(\delta_1 - \delta_t)],$$

or

$$\text{cov}\,(N_1, N_j) \cong \frac{1}{\omega^2}\,[q_1 q_j \sigma_\omega^2 - q_1 \sigma_1^2]$$
$$\cong \frac{1}{(1-p_1)^2 \tau\omega}\,[p_1 q_j \sigma_\omega^2 - p_1 \sigma_1^2]. \qquad (8.18)$$

4. Null Correlations for the Remaining-Space and Niggli-Number Correlations

From equation (8.17) we have

$$\text{var}\,(N_1) \cong \frac{1}{(1-p_1)^4}\,\text{var}\,(V_1), \qquad (8.19)$$

and from equation (8.18), similarly, that

$$\text{cov}\,(N_1, N_2) \cong \frac{1}{(1-p_1)^2}\,\text{cov}\,(V_1, V_j). \qquad (8.20)$$

Since, as already shown, var (N_j) = var (V_j) for $j \neq 1$, it follows immediately that, to first-order approximation,

$$\rho_{N_1 N_j} = \rho_{V_1 V_j}, \qquad (8.21)$$

that is, closure correlations between Niggli si and the other oxides subjected to the Niggli transformation will be the same as those between silica and the remaining-space variables resulting from elimination of silica. These are the relations most commonly shown in petrographic variation diagrams, but other relations, those between variables i and j, $1 < j < k$, are often of some interest, and it is worth stressing that on all of them the effect of both transformations, again identical, is to drastically increase the initial closure effect.

The remaining-space transformation obviously generates a new closed array. $(V_2, V_3, \ldots, V_m)$, containing one less variable than **Y**. Between each pair of major variables in this new closed array the closure correlation must be considerably stronger than that between Y's of similar subscript. For $1 < j < k$ the relation between N_j and N_k will be precisely that between V_j and V_k; they are in fact the same variables. And in the Niggli numbers proper the increase in negative correlation generated by the elimination of silica will be further reinforced by the use of sums rather than individual variables; instead of separate variables for Na_2O and K_2O, for instance, the Niggli number representing alkalies is alk = $(Na_2O + K_2O)/$

$(1 - SiO_2)$, and instead of separate variables for FeO, Fe_2O_3, and MgO the Niggli system uses the quantity $(FeO + 2Fe_2O_3 + MgO)/(1 - SiO_2)$. These summations introduce a further reduction in the number of variables, and, inevitably, a considerable further increase in the negative correlation among the survivors, quite irrespective of any substantive petrographic or geochemical associations between them.

Unfortunately, this comparison of closure correlation between remaining-space and Niggli-number variables can hardly be subjected to a realistic test. As already indicated, the whole argument presumes that the parameters of the parent open array, **X**, of our null model can in fact be estimated from the sample statistics. This is rarely possible for sample arrays described by means of Niggli (or Harker) variation diagrams, for σ calculated from such arrays nearly always contains negative elements. These must be eliminated by other transformations before the correlations expected under either of the transformations discussed in this chapter can be calculated. Transformations designed to eliminate negative elements from σ are discussed in the next chapter. Their application so modifies the raw data as to render calculation of both remaining-space variables and Niggli numbers unnecessary and that of the latter virtually impossible.

9 Harker and Related Variation Diagrams

The testing procedure outlined in the first section of chapter 6 presumes that σ, the vector of open variances computed in the fashion described in chapter 5, contains no negative elements. Like all variances, these are by definition positive. Nevertheless, negative values for one or more of them may sometimes emerge as solutions of equations (5.6) or (5.15). These always make further testing impossible but their interpretation in other respects depends somewhat on the circumstances in which they occur. An open variance negative by an amount greater than can reasonably be attributed to experimental error always means that the sample could not have been drawn from a **Y** formed by closure of an **X** in which $\sigma_{ij} = 0$ for all $i \neq j$, since such an **X** evidently does not exist.

If, in a particular context, the failure to obtain a σ free of negative elements is rare or occasional, it is probably quite safely interpreted as a rather rough group test indicating that the sample correlations as a whole may not be attributed solely to closure, even though no individual correlation or subset of correlations can be tested. If failure to obtain a σ free of negative elements is the rule rather than the exception, however, the sensible interpretation would appear to be that the null model is poorly chosen.

This seems to be the situation with regard to correlations in the suites of analyses of subalkaline volcanics ordinarily pictured in Harker diagrams. From the journal article in which it was announced (Chayes and Kruskal), as from the description given in this book, it would certainly be legitimate to assume that for the test of interdependence in such a suite of analyses the appropriate variables would be the weight or molar proportions of the various essential oxides in each specimen. To date, however, only 2 of the open variance vectors computed from suites of this kind have been free of negative elements.

1. Factors Controlling the Sign of σ_i^2 in Harker and Similar Arrays

In most suites of this type the mean and variance of silica are very much larger than those of any other variable. In fact, the joint condition $\bar{y}_1 > \Sigma_2 \bar{y}_k$, $s_1^2 > \Sigma_2 s_k{}^2$, where subscript 1 denotes silica, holds in all so far examined. We show first that under this condition $\sigma_t{}^2$, the sum of the open variances, will always be positive, and then identify the variables for which $\sigma_i{}^2$ is likely to be negative.

From equation (5. 5) we have that

$$\sigma_t{}^2 \cong \frac{\sum_{i=1}^{m} \left(\frac{s_i{}^2}{1 - 2\bar{y}_i}\right)}{\sum_{i=1}^{m} \left(\frac{\bar{y}_i(1 - \bar{y}_i)}{1 - 2\bar{y}_i}\right)}, \tag{9.1}$$

where we ignore the scaling factor, τ, and replace p_i by its sample estimate, $\bar{y}_i$.

From equation (5. 4) we may also write the ith element of $\boldsymbol{\sigma}$ as

$$\left.\begin{aligned} \sigma_i{}^2 &= \frac{1}{1 - 2\bar{y}_i}\,(s_i{}^2 - \bar{y}_i{}^2\sigma_t{}^2), \\ &= \frac{\bar{y}_i{}^2}{1 - 2\bar{y}_i}\,(C_i{}^2 - \sigma_t{}^2), \end{aligned}\right\}\ \bar{y}_i \neq 1/2 \tag{9.2}$$

where $C_i = s_i/\bar{y}_i$, the coefficient of variation of the ith variable. From this it is evident that if $\sigma_t{}^2 > 0$ we could easily determine which if any elements of a particular σ will be negative. Now equations (5. 7-5. 9) and related discussion establish that the denominator of the right side of equation (9. 1) is always negative if any $\bar{y}_i > 0.5$. Since this is always true if, as announced, $\bar{y}_1 > \Sigma_2 \bar{y}_i$, the sign of $\sigma_t{}^2$ then depends on the sign of the numerator of equation (9. 1), or

$$S = \frac{s_1^2}{1 - 2\bar{y}_1} + \sum_{i=2}^{m} \left(\frac{s_i{}^2}{1 - 2\bar{y}_i}\right). \tag{9.3}$$

But since we also have $s_1^2 > s_i{}^2$ for all $i > 1$, we may replace equation (9. 3) by the inequality

$$S < s_1^2 \sum_{i=1}^{m} \left(\frac{1}{1 - 2\bar{y}_i}\right), \tag{9.4}$$

and if $m > 2$ and some $\bar{y}_i > 0.5$ the sum on the right side of equation (9. 4) will always be negative. Under the specified conditions, accordingly, $\sigma_t{}^2$ is always positive since it is the ratio of two negative numbers.

Given the assurance that $\sigma_t^2 > 0$, it is clear from inspection that the bracketed term on the right side of equation (9. 2) controls the sign of σ_i^2. If $\bar{y}_i > 0.5$ the estimate of σ_i^2 will be negative unless the bracketed term is negative, while if $\bar{y}_i < 0.5$ it will be negative unless the bracketed term is positive. Now the bracketed term will be negative whenever the coefficient of variation of the ith variable, $(s_i/\bar{y}_i)$ is less than σ_t, and if at the same time $\bar{y}_i < 0.5$, σ_i^2 will be negative. In sum, if C_1 is unduly large σ_1 will be negative, while if C_i, $i \neq 1$, is unduly small, σ_i will be negative.

With the essential oxides as variables the first case is as yet unknown, but in Harker arrays the second is very common. It is worth adding, therefore, that if, for some one $i > 1$, say j, σ_j is indeed negative, it will also be true that C_j is smaller than any other C_i; similarly, if σ is negative for two such variables they will be the variables with smallest and next to smallest C, and so forth.

Although the practical effect of a negative element in σ is the same in either event—that is, it brings the testing to an abrupt close—one would like to be able to distinguish those which may arise as sampling errors from those which are to be regarded as overthrowing the null hypothesis. Recalling that $\sigma_t^2 = \Sigma(\sigma_i^2)$, we may write equation (9. 2) as

$$(1 - \bar{y}_i)^2\sigma_i^2 + \bar{y}_i^2 \sum_{j \neq i} (\sigma_j^2) = s_i^2, \tag{9. 5}$$

from which it is clear that $\sum_{j \neq i}(\sigma_j^2) > 0$ if $\sigma_i^2 < 0$, and also that, if σ_i^2 is negative, s_i^2 is too small. Since σ_i^2 ought to be non-negative, the smallest permissible value of s_i^2, the other sample variances and all the sample means remaining unchanged, is

$$s_i^2 \text{ (min.)} = \bar{y}_i^2 \sum_{j \neq i} (\sigma_j^2). \tag{9. 6}$$

If s_i^2 (min.) is greater than the upper bound of the α confidence interval about s_i^2, we conclude that, at the $\alpha/2$ level, the negative value of σ_i^2 is not to be dismissed as a consequence of sampling variance. (For construction of confidence limits about an observed variance see, for example, Hoel, p. 270.)

The test is far from optimum, for it assumes, in effect, that all the error is contained in one of the variances, whereas the relation between variances and covariances in closed arrays is such that an underestimate of one variance will necessarily be accompanied by underestimates of one or more of the others; if the test indicates that a particular s_i^2 is small (i.e., that σ_i^2 is negative) by an amount significant at the $\alpha/2$ level, the probability of an s_i^2 this small is probably considerably less than $\alpha/2$.

Exercises

9. 1 Show that, because of the closure constraint, $\sum_{j \neq i}(s_j^2)$ will tend to be too small if s_i^2 is an underestimate, and that $\sum_{j \neq i}(\sigma_j^2)$ will tend to be an underestimate if $\sigma_i^2 < 0$.

Hints: (a) For each variable in a closed array $\sum_{j \neq 1} s_{ij} = 0$, where s_{ij} is a variance if $i = j$ and a covariance if $i \neq j$. On the average, then, if s_{ii} is an underestimate, the sum of the covariances of u_i will be too large, and this overestimate will carry over, with opposite sign, into the estimates of s_{jj}, s_{kk}, and so on. (b) From equation (9. 5) it is evident that a gross underestimate of s_i^2 will lead to $\sigma_i^2 < 0$, and from equation (9. 1) that σ_t^2 will be small if, on the average, the elements of **s** are small. Note that overestimation of a covariance will tend to have the same effect as underestimation of a variance.

9. 2 Show that a closed "variable" which is constant—that is, has zero variance—cannot be generated by the closure of a set of uncorrelated open variables. Suggestion: Since $C_j^2 = 0$ if $s_j^2 = 0$, this curious property of our null model may be established from equation (9. 2). Given that $\sigma_t^2 > 0$ and $s_j^2 = 0$, equation (9. 2) requires that $\sigma_j^2 < 0$ if $\bar{y}_j < \frac{1}{2}$ and that $\sigma_j^2 > \sigma_t^2$ if $\bar{y}_j > \frac{1}{2}$, the latter result of course implying that $\sigma_i^2 < 0$ for some $i \neq j$. In neither case is σ free of negative elements, and this completes the proof except for the special case discussed in exercise 9. 3, below.

9. 3 Show that $\sigma_j^2 < 0$ if $s_j^2 = 0$ and $\bar{y}_j$ is exactly $\frac{1}{2}$. Suggestion: Let $s_j^2 = 0$ in equation (5. 11), recalling that $m > 2$ and that the summation does not include j. It follows at once that $\sigma_i^2 < 0$. Thus, combining the results of exercises 9. 2 and 9. 3, σ will not be free of negative elements unless all elements of **s** are positive.

9. 4 The inverse of the relation examined in the preceding two examples is false. By substitution in equation (4. 8) show that if $\sigma_j^2 = 0$ it will nevertheless be true that var $(Y_j) > 0$. Thus, there may be constants among the underlying open variables, but not among the theoretical or observed closed ones.

2. Negative Elements in σ Computed from Some Harker Arrays

The incidence of negative elements in σ calculated from each of sixteen sets of analyses used as the basis for a published diagram is shown in table 9. 1. In two of the examples σ contains no negative elements, in six it contains one, and in eight there are two. Where

Table 9.1 Incidence of Negative Elements in σ Calculated from Some Suites of Subalkaline Volcanics

Locality	Reference	Signs of Open Variances of Al_2O_3	Na_2O
Adak and Kanaga	Byers 1961	−	−
Katmai pumices	Fenner 1926	−	−
Crater Lake:	Williams 1942		
all analyses		−	−
pumices and dacites		−	+
basalts and andesites		−	−
High Cascades	Thayer 1937	−	+
Lassen Peak	Williams 1932	−	−
Medicine Lake:	Anderson 1941		
all analyses		−	−
rhyolites and dacites		−	−
basalts and andesites		−	+
Vicinity of Parícutin	Williams 1950	−	+
Parícutin volcano	Wilcox 1954	−	−
Taupo:	Steiner 1958		
ignimbrite and rhyolite		−	+
basalt and andesite		+	+
Nasu volcano	Kawano, et al. 1961	+	+
Iwaki volcano	Kawano, et al. 1961	−	+

there is only one negative element in σ, it is without exception the variance of the open variable equivalent to Al_2O_3; where there are two negative elements in σ they are always the variances of the open equivalents of Al_2O_3 and Na_2O.

To date, neither a σ with any other negative element nor a σ with more than two negative elements has been found. In twenty of the twenty-five examples the smallest coefficient of variation for variables with $\bar{y} < 1/2$ is that of Al_2O_3, and the next smallest is that of Na_2O; in the remaining five, including two of those in which σ contains no negative element, this order is reversed. This is about as complete a confirmation of the argument of the preceding section as could be desired.

3. The Elimination of Negative Elements in σ[1]

It has already been suggested that if negative elements in **σ** occur rarely and sporadically in a particular context, it is probably quite safe to argue that they provide sufficient basis for abandoning the null hypothesis, even though their presence makes any test of specific correlations impossible. If, on the other hand, negative elements are common in σ computed from a particular type of data, and perhaps especially if, as here, they tend to occur repeatedly for the same variable(s), extensive revision of the analytical approach would appear to be unavoidable. Our first concern is with the testing of individual correlations and an analytical program that frequently makes such testing impossible is obviously unrealistic.

The requirements that $p_i = \bar{y}_i$ and var $(Y_i) = s_i^2$ for all i would appear to be indispensable, but the requirement that $\sigma_{ij} = 0$ for all $i \neq j$ seems unnecessarily restrictive and probably could be relaxed; in order to accomplish this we could of course add covariances to the vectors **s** and σ and corresponding terms to the coefficient matrix, **p**, of chapter 5. If the **X** generated from the sample statistics in this fashion is free of negative elements in σ and no σ_{ij} in it exceeds $\sigma_i\sigma_j$ in absolute value, then the means, variances, and assigned covariances of **Y** will agree with those of the sample. The remaining covariances in **Y** will reflect only the effect of closure on **X**, however, and ρ formed from any one of them may be used as a null value against which to test the analogous **r**. Even the introduction of a single nonzero covariance adds considerably to the algebra, but the interested reader who has followed the argument of chapter 5 and section 5 of chapter 2 should be able to show that, for the case of $\rho_{12} \neq 0$,

$$\text{var}\ (Y_i) = p_i^2\sigma_+^2 + (1 - 2p_i)\sigma_i^2 + \theta_v\sigma_{12}, \tag{9.7}$$

where $\sigma_+^2 = \sum_{k=1}^{m} \sigma_k^2 = \sigma_t^2 - 2\sigma_{12}$

and

$$\theta_v = \begin{cases} -2p_i(1 - p_i), & i \leqslant 2 \\ 2p_i^2, & i > 2. \end{cases}$$

In analogous fashion,

$$\text{cov}\ (Y_i, Y_j) \cong p_ip_j\sigma_+^2 - p_j\sigma_i^2 - p_i\sigma_j^2 + \theta_c\sigma_{12}, \tag{9.8}$$

---

[1] This section treats a special topic and may be omitted on a first reading.

where

$$\theta_c = \begin{cases} 1 - p_i - p_j + 2p_ip_j, & i + j = 3 \\ -p_j(1 - p_i), & i \leqslant 2 < j \\ 2p_ip_j, & i, j > 2. \end{cases}$$

Equating the right side of each of the m equations of type (9. 7) to the appropriate sample variance, $s_i{}^2$, and cov (Y_1, Y_2) from equation (9. 8) to the sample covariance, s_{12}, we have, in matrix notation, that

$$\mathbf{B\sigma} = \mathbf{s} \tag{9. 9}$$

where

$$\sigma = (\sigma_1^2, \sigma_2^2, \ldots, \sigma_m^2, \sigma_{12}),$$

$$\mathbf{s} = (s_1^2, s_2^2, \ldots, s_m^2, s_{12}),$$

and

$$\mathbf{B} = \begin{bmatrix} (1-p)_1^2 & p_1^2 & p_1^2 & \cdots & p_1^2 & -2p_1(1-p_1) \\ p_2^2 & (1-p_2)^2 & p_2^2 & \cdots & p_2^2 & -2p_2(1-p_2) \\ p_3^2 & p_3^2 & (1-p_3)^2 & \cdots & p_3^2 & 2p_3^2 \\ \vdots & \vdots & \vdots & & \vdots & \vdots \\ p_m^2 & p_m^2 & p_m^2 & \cdots & (1-p_m)^2 & 2p_m^2 \\ -p_2(1-p_1) & -p_1(1-p_2) & p_1p_2 & \cdots & p_1p_2 & (1-p_1-p_2+2p_1p_2) \end{bmatrix}$$

B is the analogue of **P** in chapter 5; indeed, it is simply **P** augmented by a row and column all elements of which are sums or products of p's. Premultiplying equation (9. 9) by $\mathbf{B}^{-1}$ gives, as before,

$$\sigma = \mathbf{B}^{-1}\mathbf{s}. \tag{9.10}$$

If **B** is nonsingular[2] we thus obtain from the sample statistics estimates of the variances and nonzero covariance of a hypothetical open array, **X**, which, on closure, would yield an array, **Y**, having exactly the variances and the correlation (1, 2) observed in the sample. The other correlations in **Y**, however, will have been generated entirely by the closure of **X** and may be used as null values against which to test the significance of the analogous sample correlations.

The required null values of these correlations are the appropriate evaluations of $(p_i p_j \sigma_\ddagger^2 - p_j \sigma_i{}^2 - p_i \sigma_j{}^2 + \theta_c \sigma_{12})/s_i s_j$ in which p_i, p_j are replaced by the sample means $(\bar{y}_i, \bar{y}_j)$, θ_c is chosen as in equation (9. 8), s_i and s_j are the sample standard deviations, and $\sigma_\ddagger^2, \sigma_i$, σ_j, and σ_{12} are obtained from equation (9. 10). The test is then of the probability that some one of the observed correlations other than r_{12} might have resulted from random sampling of a **Y** array with parent means and variances equal to those of the sample and $\text{cov}(Y_1, Y_2) = s_{12}$, but all other covariances in **Y** generated entirely by closure of **X**. Additional nonzero covariances may be introduced in **X** in similar fashion, but the algebraic complication soon becomes forbidding.

Although at first glance the new null model may seem quite as easy to think about as the old one, relaxation of the requirement that $\sigma_{ij} = 0$ for all $i \neq j$ may be both more complicated and less revealing than it appears. Let us suppose, for instance, that we are interested in only one covariance, say σ_{kp}, and that we decide to choose **X** so that all other covariances in **Y** are the same as those in the sample. If the calculation is successful it leads to a situation in which the null hypothesis can always be tested but never rejected, really no improvement over a null model that precludes testing altogether! This inability to reject the null hypothesis is a direct consequence of the closure constraint, for a successful calculation will lead to a **Y** in which both the variance of Y_k and its covariances with all other variables save Y_p are identical with the analogous sample statistics. Both **Y** and the sample array being closed, however, we also have

---

[2] By an argument of the type used to show that matrix **P** of chapter 5 is nonsingular, W. Kruskal has recently shown that direct solutions for the elements of σ as defined here are in principle always obtainable, so that **B** is also nonsingular. His result will be published separately.

that var $(Y_k) + \sum_{j \neq k} \text{cov}(Y_j, Y_k) = s_k^2 + \sum_{j \neq k} s_{jk} = 0$, so that $\text{cov}(Y_k, Y_p) = s_{kp}$. Accordingly, $\rho_{kp} = r_{kp}$, and the null hypothesis cannot be rejected.

Nor is it to be supposed that a null model in which the open covariances of some particular variable are set equal to zero but all others get the values necessary to produce agreement between those of the sample and **Y** is necessarily an improvement. Such an arrangement seems very appealing in connection with the Harker correlations. Since only correlations of other elements with silica are to be tested, it might well seem reasonable to require that the open covariances of silica with each other variable be equal to zero while allowing each of the other σ_{jk}'s to take on the value required to make $\text{cov}(Y_j, Y_k) = s_{jk}$. Unfortunately, this scheme leads to exactly the same impasse as the model designed to test only one correlation, though by a somewhat more circuitous route. If open covariances of the type σ_{ij}, $j \neq i$, are set equal to zero, then only one term of the ith row of the covariance matrix of Y, the variance, is taken directly from the sample. In each other row of the matrix, however, only one term, the covariance in i, is not taken directly from the sample, and as the sum of each such row—in **Y** as in the sample—is zero, the covariance terms in i must also be identical. Hence, as before, $\rho_{ij} = r_{ij}$ and the null hypothesis can never be rejected.

The minimum requirement for a nontrivial test would appear to be that all open covariances of one variable and at least two open covariances of each other variable be zero, for then the covariances of Y_i could not be obtained by difference and would in general differ from those observed. Presumably, the choice of this second zero open covariance for each variable should be dictated by substantive considerations, but it is difficult to think of petrographic hypotheses sufficiently detailed to be of much use in this connection. A simpler scheme, one that would eliminate negative elements in $\boldsymbol{\sigma}$ without specifying any departure from zero covariance in **X**, would be preferable.

If we are to retain the zero covariance of the original null model, however, we are essentially confined to redefining the variables in such fashion as to increase the variances or decrease the means of those whose open variances are negative. These redefinitions will of course also require substantive justification, but here we may sometimes be on fairly firm ground. Before presenting a rather detailed example, we examine the algebraic effects of linear redefinition of variables on the elements of **s** and $\boldsymbol{\sigma}$, and show that the correlations between the new variables are completely implicit in sample statistics computed from the raw data.

4. Modifying Sample Variables by Linear Combination

In the preceding section we sought to modify **X** by relaxing the restriction that $\sigma_{ij} = 0$ for all $i \neq j$, and concluded that some simpler method of eliminating negative elements in σ would be preferable. Any other change in **X** will imply either further changes in the relationship of the sample to **Y**, or a change in the sample variables themselves. The first of these alternatives shifts the whole rationale of the testing procedure; the second does not. In this section we, therefore, examine the effect of transforming the sample, **U**, to, say, **V**, by linear combinations. Such transformations are common in petrography and mineralogy—every normative calculation is an example—and it is always true that the means, variances, and covariances of **V** are completely implicit in those of **U**. The variety of linear combinations is almost unlimited, and we consider only the type ordinarily encountered in normative calculations, in which **U** is initially closed and a materials balance—of weights, molar proportions, or relative numbers of atoms—must be maintained.

Suppose, for instance, that we wish to increment U_j by a multiple of itself at the expense of U_i, a procedure common in normative calculations. The full transformation is

$$V_i = U_i - \lambda U_j,$$
$$V_j = (1 + \lambda)U_j, \qquad (9.11)$$

and

$$V_k = U_k, \qquad k \neq i \text{ or } j.$$

The means and variances of the V_k and their covariances with each other are exactly those of the corresponding U_k. For V_i, however, we have that

$$\bar{v}_i = \bar{u}_i - \lambda \bar{u}_j, \qquad (9.12)$$

so that an individual deviation is

$$d_i = V_i - \bar{v}_i = \delta_i - \lambda \delta_j, \qquad (9.13)$$

and its square is

$$d_i^2 = \delta_i^2 + \lambda^2 \delta_j^2 - 2\lambda \delta_i \delta_j.$$

Summing the d_i^2's and dividing by one less than their number,

$$s_{v_i}^2 = s_i^2 + \lambda^2 s_j^2 - 2\lambda s_i s_j r_{ij}, \qquad (9.14)$$

which gives the variance of V_i as a function of the statistics of U_i and U_j. In similar fashion, the mean of V_j is

$$\bar{v}_j = (1 + \lambda)\bar{u}_j, \qquad (9.15)$$

and its variance is

$$s_{V_j}{}^2 = (1 + \lambda)^2 s_j{}^2. \tag{9.16}$$

Further, the covariance of V_i with V_j is

$$\sum (d_i d_j)/(n-1) = \frac{1}{n-1} \sum [(1+\lambda)\delta_i\delta_j - \lambda(1+\lambda)\delta_j{}^2],$$

or

$$s_{V_iV_j} = (1+\lambda)(s_i s_j r_{ij} - \lambda s_j{}^2) \tag{9.17}$$

and the correlation of V_i with V_j is

$$r_{V_iV_j} = \frac{s_{V_iV_j}}{s_{V_i} s_{V_j}} = \frac{s_i r_{ij} - \lambda s_j}{(s_i{}^2 + \lambda^2 s_j{}^2 - 2\lambda s_i s_j r_{ij})^{1/2}}. \tag{9.18}$$

The correlations between V_i or V_j and any V_k can of course be obtained by analogous procedures, so that the sample statistics for **V** may be found from those already calculated for **U**. (In small samples it will usually be quicker to transform **U** to **V** and calculate the statistics of **V** directly, but in large samples the indirect solu tion may save much copying.) Recalling that we resorted to the transformation because of negative elements in the $\boldsymbol{\sigma}$ computed directly from **U**, we note that (1) the coefficient of variation of V_j is the same as that of U_j, but (2) the mean of V_i is always smaller than the mean of U_i, and the variance of V_i will be larger than that of U_i providing only that $(\lambda s_j - 2 s_i r_{ij}) > 0$, so the coefficient of variation of V_i will often be larger than that of U_i, and if $r_{ij} < 0$ it will certainly be so.

In consequence, if the negative element in $\boldsymbol{\sigma}$ is σ_j, the transformation described above will probably be of no avail, but if $\sigma_i < 0$ it may lead to a $\boldsymbol{\sigma}$ (calculated from **V**) free of negative elements. If both σ_i and σ_j are negative, this transformation may be useful for its effect on V_i but will nearly always have to be supplemented by additional linear combinations which either increase the coefficient of variation for variable j relative to $\sigma_t{}^2$ or decrease it for some variable or variables other than i and j. The calculations will now be more involved, but it will still be true that every mean, variance, and covariance of **V** can be found from **U**, and it will usually be possible to decide in advance whether the transformation will shift the coefficients of variation in the desired direction.

Exercises

9.5 U_j is increased by a multiple of itself, say κ, at the expense of U_i and by another multiple of itself, say λ, at the expense of

U_k, that is, $V_j = (1 + \kappa + \lambda)U_j$. Show that the coefficient of variation of V_j is the same as that of U_j.

9.6 Under the conditions given in exercise 9.5 show that the covariance of V_i with V_k is $s_{ik} - \lambda s_{ij} - \kappa s_{jk} + \kappa\lambda s_j^2$.

9.7 U_j and U_k are increased by multiples of themselves, say κ and λ, at the expense of U_i. Show that $\bar{v}_i = \bar{u}_i - \kappa\bar{u}_j - \lambda\bar{u}_k$ and $s_{v_i}{}^2 = s_i{}^2 + \kappa^2 s_j{}^2 + \lambda^2 s_k{}^2 - 2(\kappa s_{ij} + \lambda s_{ik}) + \kappa\lambda s_{jk}$.

Note again that, as in the simpler example used in the text, $\bar{v}_i < \bar{u}_i$ while $s_{v_i}{}^2$ may or may not be larger than $s_i{}^2$. The overall effect would usually be to enlarge the coefficient of variation of V_i relative to U_i and thus reduce the risk that $\sigma_i{}^2 < 0$ in σ calculated from **V**. This transformation is of central importance in the following discussion.

5. Some Linear Transformations Applicable to the Harker Array

In this section, as previously, a linear combination involving two or more of the elements of **U** is formally regarded as a transformation of all, that is, **V** = **KU**, where, if $U_k = V_k$, the kth element of the kth row of **K** is unity and all other elements in this row are zero. For ease of reference, however, we replace alphabetical by numerical subscripts. Specifically, the oxides reported in each analysis are arranged and referred to in the order (1) SiO_2, (2) Al_2O_3, (3) Fe_2O_3, (4) FeO, (5) MgO, (6) CaO, (7) Na_2O, (8) K_2O, (9) TiO_2, (10) Rest. Thus "variable 7" refers, as required, to the seventh element of **U**, of **X**, $\boldsymbol{\mu}$, and $\boldsymbol{\sigma}$ obtained or approximated from the assemblage of sample vectors **U**, of **V** calculated from **U**, and of **X**, μ, and σ approximated from the assemblage of transformed sample vectors **V**.

In this notation, our problem is that in subalkaline Harker arrays σ_2^2 and σ_7^2 computed from the array of sample vectors $\mathbf{U}_h$, h = 1, 2, . . . , n, are nearly always negative, so that the parent variances and covariances of **Y** are undefined. We seek linear transformations of the type $\mathbf{V}_h = \mathbf{KU}_h$, such that $\boldsymbol{\sigma}$ computed from **V** will be free of negative elements, each $\mathbf{V}_h$ of course being subject to the condition that the original materials balance is maintained, that is, the sum of the elements of $\mathbf{V}_h$ is the same as that of $\mathbf{U}_h$.

If we are to use the testing procedure described in chapter 6 we must have a $\boldsymbol{\sigma}$ free of negative elements. This is not possible with the conventional choice of variables, and the practical issue is not whether but how these variables are to be redefined. From the work of the preceding sections of this chapter it is clear that only transformations which increase the coefficients of variation of variables 2 and 7 or materially reduce the variance of variable 1 need be considered. In discussing the algebra of the linear combination technique, however, it is easy to lose sight of the fact that no transformation should be used unless there is substantive justification for it. That a transformation is needed and this particular one works—that is, that **σ** computed from **U** contains negative elements and σ computed from **V** does not—is not enough.

Our null model presumes zero covariance between the elements of **X** and, for the present at least, we have agreed not to relax this restriction. It is well known, however, that in the solid state there are strong nonprobabilistic associations between some of the elements of **U**, or, more precisely, between some parts of some of the elements of **U**. In eucrystalline rocks of the sort under discussion, for instance, pyroxenes and amphiboles are subalkaline, feldspathoids are lacking, and the bulk of the alkali content enters feldspar in which the normative ternary molar ratio $R_2O \cdot Al_2O_3 \cdot SiO_2$ is 1 : 1 : 6. A small proportion of the alkalies not contained in feldspar is contained in mica, in which there is again a very strong association of R_2O with Al_2O_3. The 1 : 1 : 6 molar ratio of alkali feldspar has been shown to survive solution and transport by water vapor at high pressure (Morey and Chen 1955), and it seems reasonable to suspect that a linkage so persistent in the solid and vapor states would also persist in rock liquids.

If we knew the partition of alkalies between feldspar and mica we could devise a linear transformation such that V_1 and V_2 would be the amounts of SiO_2 and Al_2O_3 remaining after fixation of alkalies in feldspar and mica, and V_7 and V_8 the amounts of (ab + Na-mica) and (or + K-mica) respectively. Alternatively, we could let V_7 represent all alkali feldspar and V_8 all (aluminous) mica. Unfortunately, the partition function is unknown, and we are obliged to guess.

In what terms shall we guess? The data are originally in units of weight percent, but many petrologists feel that molar proportions would be preferable and they are often used in projections even though the analyses themselves are posted in weight percent. It is often considered that the coordinates of a Harker diagram must be weight percents, but in the original variation diagram of this sort (Iddings 1892) the coordinates are molar proportions. Devotees of weight percent units often argue that the weights of the oxides are what the analyst actually observes, but this is incorrect. In conventional analysis the final arbiter is indeed the balance but the weighings are rarely of oxides; most of the weight percentages listed in the conventional analysis are interpretations based on the concept of molecular or combining weights. The petrologist to whom molecular proportions seem more "natural" or less "arbitrary" often points out that they greatly simplify calculations. This is true and important if the calculations are to be done by hand, less true and rather unimportant if a desk calculator is used, and untrue if the work is to be done on a programmed computer. All of the computations reported below have been carried out on six well-known North

American suites with **U** in both units, but the detailed discussion assumes throughout that **U** is in molar proportions because the effect of transformation on the distribution of negative elements in σ seems much more consistent. Results for **U** in weight percent are given in tabular form, however, and are briefly compared with those for **U** in molar percent at the end of the section.[3]

Returning now to the question of which linear combinations may be expected to yield the desired results, it will be recalled that in rocks of this sort the "transformation" implicit in the CIPW normative conventions concerning alkalies is simply the conversion of all K_2O to or and all Na_2O to ab. With **U** in molar proportions, this may be written

$$V_1 = U_1 - 6U_7 - 6U_8,$$
$$V_2 = U_2 - U_7 - U_8,$$
$$V_7 = 8U_7,$$
$$V_8 = 8U_8,$$

and

$$V_k = U_k, \text{ for } k \neq 1, 2, 7, \text{ or } 8.$$

As shown by the sixth line of table 9. 2, however, this leads to an overcompensation; the original negative elements of σ, 2 and 7, are eliminated for each set of analyses, but in three of the sets a new negative element appears, for the open equivalent of silica.

The conversion of all alkali to feldspar being unsuccessful and the partition function of alkalies between (ab + or) and other components being unknown, the next choice would appear to be the conversion of either Na_2O to ab or K_2O to or. The normative conventions would dictate the latter, but the second and third lines of table 9. 2A show that neither is successful. In all six test groups both negative variance elements survive the or transformation, and in each the negative variance of the open equivalent of Al_2O_3 survives the ab transformation. The shifts in means and variances are thus too great if both alkalies are subjected to "feldspar" transformations and not great enough if only one is treated in this fashion. If, in addition to the or transformation, Na_2O is complexed as $NaAlO_2$, the negative variance elements for the open equivalent of Al_2O_3 vanish, but those for the open equivalent of Na_2O persist in five of the six test groups. On the other hand, if, in addition to the ab transforma-

[3] Since the variable "Rest" has no unique molecular weight, the conversion from weight to molar proportions involves elimination of minor constituents, chiefly H_2O, and, of course, a new closure.

Table 9.2 Negative Elements in σ Surviving Various Linear Transformations of Some Well-Known Harker Arrays

Transformation	Lassen	Crater Lake	Medicine Lake	Parícutin	Adak and Kanaga	Katmai
A. Molar Proportions						
none	2, 7	2, 7	2, 7	2, 7	2	2, 7
or	2, 7	2, 7	2, 7	2, 7	2	2, 7
ab	2	2	2	2	2	2
or + $NaAlO_2$	7	7	7	7	none	7
ab + $KAlO_2$	none	none	none	none	none	none
ab + or	none	1	1	none	none	1
B. Weight Proportions						
none	2, 7	2, 7	2, 7	2, 7	2, 7	2, 7
or	none	none	1	1	7	none
ab	none	none	none	7	none	none
or + $NaAlO_2$	none	none	1	1	7	1
ab + $KAlO_2$	none	none	none	7	none	none
ab + or	none	none	7	7	7	7

tion, K_2O is complexed as $KAlO_2$, none of the open-variance vectors contains negative elements. For these data, and all other published Harker arrays on which it has been tried, the ab + $KAlO_2$ transformation is successful. The transformation may be written $\mathbf{V}_p = \mathbf{K}\mathbf{U}_p$ where $\mathbf{V}_p$ and $\mathbf{U}_p$ are as previously defined and

$$\mathbf{K} = \begin{vmatrix} 1 & 0 & 0 & 0 & 0 & 0 & -6 & 0 & 0 \\ 0 & 1 & 0 & 0 & 0 & 0 & -1 & -1 & 0 \\ 0 & 0 & 1 & 0 & 0 & 0 & 0 & 0 & 0 \\ 0 & 0 & 0 & 1 & 0 & 0 & 0 & 0 & 0 \\ 0 & 0 & 0 & 0 & 1 & 0 & 0 & 0 & 0 \\ 0 & 0 & 0 & 0 & 0 & 1 & 0 & 0 & 0 \\ 0 & 0 & 0 & 0 & 0 & 0 & 8 & 0 & 0 \\ 0 & 0 & 0 & 0 & 0 & 0 & 0 & 2 & 0 \\ 0 & 0 & 0 & 0 & 0 & 0 & 0 & 0 & 1 \end{vmatrix} .$$

From table 9. 2B it will be noted that none of the five proposed transformations is completely successful if **U** is in weight rather than molar percent. Two, ab and (ab + $KAlO_2$), fail only for one of the six sets, so that in this respect the margin of superiority of the molar proportions is not large. As already suggested, however, the distribution of negative elements seems considerably more consistent if **U** is in molar proportions; if molar proportions are used, the or transformation eliminates none of the negative elements in $\boldsymbol{\sigma}$, the ab transformation makes σ_7^2 positive, the (or + $NaAlO_2$) transformation makes σ_2^2 positive, and the (ab + $KAlO_2$) eliminates all negative elements from $\boldsymbol{\sigma}$. In contrast to this surprisingly consistent sequence, if **U** is in weight proportions all of the transformations make σ_2^2 positive, all of them eliminate negative values for σ_7^2 as well in two of the test arrays, but there is little to choose between in their performance on the remaining four. Negative values for σ_7^2 are not completely eliminated by any of them and two introduce negative values for σ_1^2.

Exercises

9. 8 The transformation $K_2O \rightarrow KAlO_2$ is of the type described by the unnumbered display preceding equation (9. 12), above. The mean and variance of $KAlO_2$ are given by equations (9. 15) and (9. 16). If **U** is in weight percent, what is the numerical value of λ?

9. 9 The transformation $Na_2O \rightarrow$ ab is of the type described in exercise 9. 5. If **U** is in weight percent, what are the numerical values of κ and λ?

9.10 In the combined transformation ($Na_2O \rightarrow$ ab, $K_2O \rightarrow KAlO_2$), note that Al_2O_3, U_2, plays the role of U_i in exercise 9.5.

9.11 (a) With **K** as defined in this section, find cov (V_1, V_2) as a function of the sample statistics. (b) Replace each element of **K** by the value required if the calculations were to be carried out on weight percentages instead of molar percentages.

6. Testing the Significance of Correlations between Transformed Harker Variables

Following the (ab + $KAlO_2$) transformation on molar proportions, the means, variances, and "Harker" correlations of the Lassen Peak suite are as shown in table 9.3. The column headed "t" may now be found and evaluated in the fashion described in detail in chapter 6. The 0.05 and 0.01 points for t are 1.96 and 2.58; all of the "Harker" correlations save those for Al_2O_3, Fe_2O_3, and FeO are easily significant at the 0.01 level. The correlation between ab and SiO_2 is an excellent illustration of the lack of correspondence between graphical and numerical criteria of association in closed arrays; $r_{17} = 0.4888$, in absolute value the smallest but one of the observed correlations, yields the largest t value.

Table 9.3 Transformed Variables and "Harker" Correlations for Lassen Peak Volcanics (molar percent)

			Correlation with $(SiO_2)'$		
Variable	Mean	Variance	Observed (r)	Null (ρ)	|t|
1. $(SiO_2)'$	43.21	13.8692			
2. $(Al_2O_3)'$	5.72	1.8346	−0.6852	−0.4355	1.93
3. Fe_2O_3	0.79	0.2420	−0.3630	−0.1447	1.22
4. FeO	3.00	1.8029	−0.6202	−0.2522	2.43
5. MgO	6.06	8.5455	−0.7887	−0.3773	3.49
6. CaO	7.44	4.3372	−0.8541	−0.3988	4.41
7. ab	30.80	13.6636	0.4888	−0.7229	7.52
8. $KAlO_2$	2.53	0.6589	0.7565	−0.2665	6.55
9. TiO_2	0.45	0.0360	−0.6378	−0.1732	3.01

Note: $(SiO_2)'$ is the silica remaining after conversion of Na_2O to ab. $(Al_2O_3)'$ is the alumina remaining after conversion of Na_2O to ab and K_2O to $KAlO_2$.

References

Anderson, C. A. 1941. *Volcanoes of the Medicine Lake Highland, California.* Publ. Univ. Calif., Bull. Dept. Geol. Sci., vol. 25, no. 7, pp. 347-422.

Byers, F. M. 1961. Volcanic suites, Umnak and Bogoslof Islands, Aleutian Islands, Alaska. *Bull. Geol. Soc. Am.* 72: 93-128.

Chayes, F. 1949. On correlation in petrography. *J. Geol.* 57: 239-54.

———. 1962. Numerical correlation and petrographic variation. *J. Geol.* 70: 440-52.

———. 1967a. Modal composition of U.S.G.S. reference sample G-2. *Geochim. Cosmochim. Acta* 31: 463-64.

———. 1967b. On the graphical appraisal of the strength of associations in petrographic variation diagrams. In *Researches in geochemistry*, ed. P. H. Abelson, vol. 2, pp. 322-39. New York: Wiley.

Chayes, F., and Kruskal, W. 1966. An approximate statistical test for correlations between proportions. *J. Geol.* 74: 692-702.

Feller, W. 1950. *An introduction to probability theory and its applications.* New York: Wiley.

Fenner, C. N. 1926. The Katmai magmatic province. *J. Geol.* 34: 673-772.

Flight, W. 1887. *A chapter in the history of meteorites.* London: Dulan.

Gast, Paul. 1965. Terrestrial ratio of potassium to rubidium and the composition of the earth's mantle. *Science* 147: 858-60.

Goodman, L. A., and Kruskal, W. H. 1963. Measures of association for cross-classifications: III. Approximate sampling theory. *J. Am. Stat. Assoc.* 58: 310-64.

Goodman, R. 1966. *Teach yourself statistics.* London: English Universities Press.

Hart, S. R., and Nalwalk, A. J. 1970. K, Rb, Cs and Sr relationships in submarine basalts from the Puerto Rico trench. *Geochim. Cosmochim. Acta* 34: 145-55.

Hoel, P. G. 1962. *Introduction to mathematical statistics*, 3d ed. New York: Wiley.

Iddings, J. P. 1892. The origin of igneous rocks. *Bull. Phil. Soc. Wash.* 11: 89-202.

Kawano, Y.; Yagi, K.; and Aoki, K. 1961. Petrography and petrochemistry of the volcanic rocks of Quaternary volcanoes of northeastern Japan. *Sci. Repts. Tohoku Univ.* (ser. 3) 7(1): 1-46.

Kendall, M. G., and Stuart, A. 1963. *The advanced theory of statistics*, 2d ed., vol. 1. New York: Hafner.

Kermack, K. A., and Haldane, J. B. S. 1950. Organic correlation and allometry. *Biometrika* 37: 30-41.

Kruskal, W. H. 1953. On the uniqueness of the line of organic correlation. *Biometrics* 9: 47-58.

Ku, H. H. 1965. *Notes on the use of propagation of error formulas*. National Bureau of Standards Report no. 9011. Washington, D. C.

Morey, G. W., and Chen, W. T. 1955. The action of hot water on some feldspars. *Am. Mineralogist* 40: 996-1000.

Mosimann, J. 1962. On the compound multinomial distribution, the multivariate β-distribution and correlations among proportions. *Biometrika* 49: 65-82.

Pearson, K. 1896-97. On a form of spurious correlation which may arise when indices are used in the measurement of organs. *Proc. Roy. Soc.* (London) 60: 489-502.

Russell, R. D., and Farquhar, R. M. 1960. *Lead isotopes in geology*. New York: Interscience.

Snedecor, G. W. 1956. *Statistical methods*, 5th ed. Ames, Iowa: Iowa State College Press.

Steiner, A. 1958. Petrographic implications of the 1954 Ngauruhoe lava and its xenoliths. *New Zealand J. Geol. Geophys.* 1: 325-63.

Thayer, T. P. 1937. Petrology of later Tertiary and Quaternary rocks of the north-central Cascade Mountains in Oregon, with notes on similar rocks in western Nevada. *Bull. Geol. Soc. Am.* 48: 1611-52.

Wilcox, R. E. 1954. *Petrology of Parícutin volcano, Mexico.* U.S. Geological Survey Bulletin 965-C, pp. 281-365. Washington, D.C.

Williams, H. 1932. *Geology of the Lassen Volcanic National Park.* Publ. Univ. Calif., Bull. Dept. Geol. Sci., vol. 21, no. 8, pp. 195-385.

———. 1942. *The geology of Crater Lake National Park, Oregon, with a reconnaissance of the Cascade Range southward to Mount Shasta*. Carnegie Institution of Washington Publication 540. Washington, D.C.

———. 1950. *Parícutin region, Mexico*. U.S. Geological Survey Bulletin 965-B, pp. 165-279. Washington, D.C.